Advanced Manufacturing Technology for Medical Applications

Advanced Manufacturing Technology for Medical Applications

Reverse Engineering, Software Conversion and Rapid Prototyping

Edited by

Ian Gibson
Department of Mechanical Engineering
University of Hong Kong
Hong Kong

John Wiley & Sons, Ltd

Other Wiley Editorial Offices

John Wiley & Sons Inc., 111 River Street, Hoboken, NJ 07030, USA

Jossey-Bass, 989 Market Street, San Francisco, CA 94103-1741, USA

Wiley-VCH Verlag GmbH, Boschstr. 12, D-69469 Weinheim, Germany

John Wiley & Sons Australia Ltd, 42 McDougall Street, Milton, Queensland 4064, Australia

John Wiley & Sons (Asia) Pte Ltd, 2 Clementi Loop #02-01, Jin Xing Distripark, Singapore 129809

John Wiley & Sons Canada Ltd, 22 Worcester Road, Etobicoke, Ontario, Canada M9W 1L1

Wiley also publishes its books in a variety of electronic formats. Some content that appears in print may not be available in electronic books.

Library of Congress Cataloging-in-Publication Data

Advanced manufacturing technology for medical applications : reverse engineering, software conversion, and rapid prototyping / edited by Ian Gibson.
    p.   cm.
Includes bibliographical references.
ISBN 0-470-01688-4 (cloth : alk. paper)
1. Medical technology.   2. Medical innovations.   3. Manufacturing processes.   I. Gibson, Ian, 1938–
R855.3.A38   2005
658′.28—dc22

                                                                        2005021391

British Library Cataloguing in Publication Data

A catalogue record for this book is available from the British Library

ISBN-13 978-0-470-01688-6 (HB)
ISBN-10 0-470-01688-4 (HB)

Typeset in 10/12pt Times by TechBooks, New Delhi, India
Printed and bound in Great Britain by Antony Rowe Ltd, Chippenham, Wiltshire
This book is printed on acid-free paper responsibly manufactured from sustainable forestry
in which at least two trees are planted for each one used for paper production.

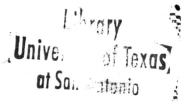

Contents

Contributors

Andrew Christensen, CEO
Medical Modeling LLC,
17301 West Colfax Avenue,
Suite 300,
Golden,
Colorado 80401, USA
Email: information@medicalmodeling.com

Dr. Denis Cormier, Associate Professor
Department of Industrial Engineering,
North Carolina State University,
Raleigh,
North Carolina 27695-7906, USA
Email: cormier@ncsu.edu

Ellen Dhoore
Materialise,
Technolgielaan 15,
3001 Leuven,
Belgium
Email: http://www.materialise.com/location/Malaysia_ENG.html

Advanced Manufacturing Technology for Medical Applications   Edited by I. Gibson
© 2006 John Wiley & Sons, Ltd.

Paul D'Urso, Consultant Neurosurgeon
Department of Neurosurgery,
Alfred Hospital,
Commercial Road,
Melbourne,
Victoria 3004,
Australia
Email: http://www.alfred.org.au/departments/neurosurgery_department.html

Dr. Wei Feng, Research Fellow
Department of Mechanical Engineering,
National University of Singapore,
Faculty of Engineering,
9 Engineering Drive 1,
Singapore 117576
Email: mpefengw@nus.edu.sg

Dr. Jerry Fuh Ying Hsi, Associate Professor
Department of Mechanical Engineering,
National University of Singapore,
Faculty of Engineering,
9 Engineering Drive 1,
Singapore 117576
Email: mpefuhyh@nus.edu.sg

Dr. Ian Gibson, Associate Professor
Department of Mechanical Engineering,
The University of Hong Kong,
Haking Wong Bdg,
Pokfulam Road,
Hong Kong
Email: igibson@hku.hk

Dr. Ola Harrysson, Assistant Professor
North Carolina State University,
Department of Industrial Engineering,
119 Park Shops,
Raleigh,
North Carolina, 2769, USA
Email: harrysson@ncsu.edu

Dr. Dietmar Hutmacher, Associate Professor
Division of Bioengineering,
National University of Singapore,
Faculty of Engineering,
9 Engineering Drive 1,
Singapore 117576
Email: biedwh@nus.edu.sg

Robert Thompson, General Manager
Anatomics Pty. Ltd,
PO Box 4012,
Eight Mile Plains,
Queensland 4113,
Australia
Email: anatomics@anatomics.com

Dr Wong Yoke-San, Associate Professor
Department of Mechanical Engineering,
National University of Singapore,
Faculty of Engineering,
9 Engineering Drive 1,
Singapore 117576
Email: mpewys@nus.edu.sg

Dr Yunfeng Zhang, Assistant Professor
Department of Mechanical Engineering,
National University of Singapore,
Faculty of Engineering,
9 Engineering Drive 1,
Singapore 117576
Email: mpezyf@nus.edu.sg

1

Rapid Prototyping for Medical Applications

Ian Gibson

1.1 Overview

While rapid prototyping (RP) technology has primarily been developed for the manufacturing industry to assist in speeding up the development of new products, its vendors and users were quick to realize the technology was also suitable for applications in the medical field. Doctors and surgeons have always been looking for better ways to describe, understand and diagnose the condition of individual patients. Diagnostic tools have become increasingly more sophisticated and the latest CT, MRI and other medical imaging technology can now present patient data in many ways and with great clarity and accuracy. There are, however, many cases where doctors or surgeons might like to have a physical model in front of them rather than have to look at images on a computer screen. Before RP, such models could only be generic and were not necessarily useful to describe an individual condition. With RP there came a way to create such physically solid models of an individual directly from the 3D data output by the medical imaging system.

RP is an emerging technology. In recent years, a number of books have been written on the subject, either describing the basic technologies from which RP machines have been constructed and/or their applications. Some of the more recently written books have even included medical applications. However, none of these books has described medical applications in any real detail or included a very wide range of applications. This book sets out to address both these points.

RP is becoming used for an increasing number of medical applications as the real benefits become more widely known and appreciated. However, many medical practitioners do not fully understand, or indeed even know about, these applications and this book aims to present these

Advanced Manufacturing Technology for Medical Applications   Edited by I. Gibson
© 2006 John Wiley & Sons, Ltd.

to those medical experts. In addition, many engineers familiar with RP would like to understand more completely where these benefits lie so that they can communicate more effectively with medics. While it is well known that RP can be of benefit, the 'actual' uses are less clear.

1.2   Workshop on Medical Applications for Reverse Engineering and Rapid Prototyping

The idea for putting together this book came to me when I was on study leave at The National University of Singapore in the second half of 2003. As a long-standing researcher into RP and co-editor of the *Rapid Prototyping Journal*, I have looked for and supported research that broadens the applications for RP. I had previously carried out a number of projects at my home University of Hong Kong, supporting surgeons with models of patient data, and wanted to explore this area during my visit to Singapore. Singapore has long been famous for advanced surgery and had recently gained much publicity and admiration for work on separating conjoined twins. During my stay in Singapore I met with a number of medics and engineers involved in these cases and was interested to see the vital role RP played in the surgical planning. I quickly learned that in fact such operations would not have been carried out had it not been possible to make these models.

Two years previously I had put together a book on research into software systems for RP. The experience of putting together this book was a very profound and fulfilling experience. The approach was to collect together a group of like-minded experts working in the area and run a book workshop. Each expert contributed a chapter to the book and the workshop allowed us all to comment and collectively tune the content. If anyone is thinking of putting together a research book, I can highly recommend this approach. It has many benefits, in addition to the resulting book itself, in terms of establishing professional and personal relationships with experts in similar fields. Having decided that a book should be written on the subject, I did not hesitate to set up a similar event in order to put it together. There were a number of criteria I insisted upon for this event:

- The workshop should have representatives from key commercial organizations involved in making medical models.
- The subjects should include research into a number of key areas: tissue engineering, surgical modelling, implant technology, materials development and different surgical applications.
- There should be input from medical practitioners.
- Representatives should be from around the world.
- There should be a mixture of research and commercial applications.

I then set about finding out who were the experts in the various fields and inviting them to Singapore. As an added incentive to the book workshop, I also organized an open workshop for outside parties in Singapore to learn from these experts. This made it possible for many of these extremely busy individuals to justify attending the event. The resulting workshops were eventually held at The National University of Singapore from 3 to 5 December 2003.

I was delighted by the response and enthusiasm of the workshop participants who have all gone on to supply chapters for this book. I hope you agree that we were able to meet all my criteria and that the collective expertise of these authors represents a significant proportion of the worldwide expertise in this field. I will allow the authors to present themselves in whatever way they wish, but it can be seen that we have representatives from three of the top (in fact probably the top three) companies that fabricate medical models. We also have two of the top research groups into tissue engineering using RP and representatives from a number of important emerging research groups in Asia, Europe and the United States. Applications described include dentistry, neurosurgery, implant design and development, maxillofacial surgery, orthopaedics and separation of conjoined twins.

I hope you agree with me that this book is indeed a mine of information on research and applications of RP for medical applications, and I also hope you gain much support for your own research and application.

1.3   Purpose of This Chapter (Overview)

Most of the chapters of this book cover the use of advanced manufacturing technology for medical applications. Most of the chapters make the assumption that the reader already knows much about the technology used. For many readers we anticipate that this will indeed be the case. However, there will be a number of readers, particularly those with a medical background, who know little or nothing about this technology. This introductory chapter will therefore aim to assist the reader in understanding the basics of RP technology. Where other technologies are involved, for example in reverse engineering and software systems, the authors of the relevant chapters have taken efforts to ensure that technology and terminologies are explained in a clear and concise manner. However, nearly all the chapters include reference to RP technologies of different kinds, and so it is worthwhile devoting some time and effort to explaining these technologies in this single introductory chapter, rather than have the authors repeat themselves too often. This also permits the authors to concentrate on the more important aspects of their work rather than wasting pages to explain basic terms to the reader.

This chapter therefore aims to describe a number of key rapid prototyping technologies. There are many RP technologies commercially available, and it is not within the scope of such a short chapter to describe them all, or in great detail. If readers require more detail, they are advised to refer to one of the numerous excellent books that are dedicated to describing these technologies. A bibliography of the more popular of these books can be found at the end of this chapter. This chapter sets out, if you will pardon the pun, to cover the bare bones of the subject. It also focuses on those technologies that are best suited to medical applications rather than listing all the ever increasing number of available technologies. However, before listing the different machines that are capable of making physical models from medical data, it is necessary to explain the basic principles of RP.

1.4   Background on Rapid Prototyping

Rapid prototyping is a term used to represent a range of technologies that can fabricate 3D objects in a single stage, directly from their CAD descriptions. There are a number of other terms associated with RP that can be used to describe the technology:

- Freeform fabrication. This emphasizes the fact that RP is largely 'geometrically independent' in that any increase in the complexity of form does not necessarily make it more difficult to build.
- Automated fabrication. This links RP with other, similar technologies such as NC machining to emphasize the fact that parts are largely produced without human intervention. Since RP

replaces traditional modelmaking skills, this can be considered a huge advantage in terms of increased manufacturing speed, throughput and reduced demand on skilled labour.
- Layer-based additive manufacture. RP simplifies the complex 3D fabrication process by reducing it to a series of finite-thickness 2D forms or layers and adding them together.

The RP term itself has been the subject of much discussion and controversy. RP is also used to describe processes in the software, business and electronics sectors. The general definition relates to being able to create objects, models or systems in a speedy manner so that further development is subsequently streamlined. RP is therefore a relatively ambiguous term that is also linked to rapid product development. However, RP was the first term used to popularize this technology and as such it has stuck.

The final definition of layer-based manufacture is the key to how RP really works. Models are created by bonding layers of material together. If the layers are sufficiently thin, then the models will closely approximate the original intended design. Most RP processes use layer thicknesses of the order of 0.1 mm, and this seems to be sufficient to suit many applications. RP is therefore becoming a well-accepted technology in the manufacturing sector, with many different industries (e.g. kitchenware, electronics, toymaking, jewellery, automotive and aerospace, etc.) making use of its capabilities. The term 'additive' manufacture also distinguishes RP from NC machining, which uses a stock material and removes, or subtracts, material to reveal the final shape. With RP you start with just a substrate and add material in layers until the part is completed.

People familiar with medical imaging systems should understand quite easily how the basic concept of RP works. CT and MRI imaging systems both work in a similar manner. 3D images of patient data are constructed by combining 2D slices taken from the sensor systems and interpolating between them. Generally, the slice separation is quite large and coarse in comparison with RP machines, although improved sensors and techniques are enabling models of increasing accuracy to be created. Thus, essentially, the two processes are very similar. CT and MRI combine software slices to create a 3D model, and RP takes a 3D model and reproduces it in a physical form by combining layers together. It is perhaps this synergy between the two types of system that encourages many of us to explore the use of RP for making medical models. The medical imaging systems are capable of creating virtual models, but many medical applications can further benefit from having physical models made from these data. Doctors and surgeons, as we will come to realize through the chapters of this book, are highly physical and therefore tactile people. Being able to feel a representation of the patient data in their hands can go a very long way to solving problems concerning their treatment.

There are many different ways in which the part layers can be made, and consequently there are numerous different RP machines and manufacturers. This chapter will go on to discuss some of the more popular and relevant machines that are suitable to medical applications. However, it should also be noted that RP technology has many limitations, and it is appropriate at this stage to list the most common:

- Layer thickness. A 0.1 mm layer thickness is still too thick for many applications (although probably not a concern for many medical models). The best layer thickness commonly available is around 0.02 mm, but users must be made aware that reduction in layer thickness means more expensive machines and a slower build process. All RP parts exhibit a characteristic 'stair step' texture, most evident on sloping surfaces, and therefore manual surface finishing is required for many models.

- Part accuracy. In addition to layer thickness, there are a number of other accuracy issues that affect the building of parts. In particular, one can expect an RP machine to have a minimum wall thickness for shell-type parts. Normally this will be a few tens of millimetres. Repeatability of RP processes is generally good (of the order of a few microns), but part shrinkage due to material and process constraints can lead to tolerances in the few tenths of a millimetre range.

- Part size. Geometrical independence is not strictly true in that models are restricted by the working envelope of the machine. Many RP machines are of the order of $300\,\text{mm}^3$. This means that a model of a full human skull would be difficult to make in many machines and that many medical applications may require construction of parts in sections or to scale.

- Materials. Much of the part fabrication process in RP is dependent on the ability to combine layers together. This forces severe constraints on the materials suitable for a particular process. The majority of RP processes build models from polymeric materials since this represents satisfactory material properties without the need to resort to very high temperatures and/or forces.

- Part strength. Since parts are built in layers, which are then bonded together in some way, it is likely that these bonded regions represent weaknesses in the overall structure. Even within the material range of a particular process, it is commonly found that the mechanical strength of parts made is slightly inferior to that of parts made with the same material in other manufacturing processes.

- Speed. RP is not as 'rapid' as many people realize and would like. Parts generally take from a matter of hours to perhaps a couple of days to fabricate, depending on the chosen process and size of part. While this is a significant improvement on conventional model-making approaches (with the addition of improved accuracy, material properties, etc.), there is always a demand for further increase in speed. A surgeon working on an emergency case may not be able to wait for models to be made in this time frame.

- Cost. Of course, the capacity to create models quickly, accurately and reliably using RP technology must come at a price. RP technology is still something of a novelty and machines are generally put together as small-volume production. Many of the higher-end machines are over US$200 000. Having said that, the prices of all machines are steadily coming down, and smaller machines that are focused more at the concept modelling sector, generally with limited properties, are approaching a cost similar to many high-end computer products (i.e. around US$30 000). Many applications for medical models can be addressed using these lower-cost machines.

This means that, as researchers and developers, we have plenty of work to keep us busy. For example, systems are being developed that can make parts with micron-scale layer thicknesses. Other machines are focusing on producing parts with a wider variety of materials, namely biomaterials and metals. Ink-jet printing technology is widely used in more recent RP machines, with the major benefits of controllable precision and increased speed of build. Add to this research the incremental improvements in existing technology and system integration and you have a huge amount of potential for achieving the ultimate goal of rapid manufacture.

Medical applications may be classified in terms of demand for high speed, low cost and integration with medical procedures. Accuracy is generally a low-priority issue in comparison with most engineering applications. Material properties have very special requirements related to biocompatibility and approval from medical authorities for use in operating theatres, etc.

1.5 Stereolithography and Other Resin-type Systems

The first ever commercial RP systems were resin-based systems commonly called stereolithography or SLA. The resin is a liquid photosensitive polymer that cures or hardens when exposed to ultraviolet radiation. The UV light comes from a laser, which is controlled to scan across the surface according to the cross-section of the part that corresponds to the layer. The laser penetrates into the resin for a short distance that corresponds to the layer thickness. The first layer is bonded to a platform, which is placed just below the surface of the resin container. The platform lowers by one layer thickness and the scanning is performed for the next layer. This process continues until the part has been completed.

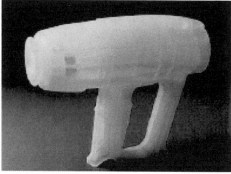

Figure   1.1 Stereolithography equipment with a typical component (courtesy of 3D Systems)

Since the surrounding uncured resin is still in liquid form, any overhanging features of the part must be supported during the build process to prevent them from collapsing under their own weight. Since the part is being built layer by layer, some regions may also be initially detached from the remainder of the part. These detached regions must also be supported to prevent them from floating away on the top of the resin. Parts must therefore be built with additional support structures. These must be removed from the part once the build process has been completed.

The accuracy of this process is generally considered to be among the best of all RP technologies, and SLA resins are generally transparent or translucent, which is a very useful characteristic as far as medical applications are concerned. Being able to look into the resin and see internal features, essentially making the bone appear transparent, can provide a surgeon with much insight into the patient's condition. One particular resin developed specifically for medical applications makes it possible selectively to colour regions within the part by overexposing these regions to UV light. Once completed, parts can be 'fixed' by applying a lacquer that is clear in visible light but blocks out UV. Since the remainder of the SLA part is transparent, the coloured regions can be used to highlight features within the part. For example, the vascular structure can be highlighted in this way and the surgeon can plan the surgical procedure such that blood vessels can be avoided.

SLA is quite an expensive technology, but there are now a number of other resin-based systems at much cheaper prices. However, the original SLA technology from 3D Systems is still widely considered to be the best example of this kind of technology.

1.6 Fused Deposition Modelling and Selective Laser Sintering

There are many ways to distinguish the different RP technologies. Most references would not choose to collect together the selective laser sintering (SLS) process with the fused deposition modelling (FDM) process. However, in relation to medical purposes, applications can be quite similar.

These technologies use heat to melt the base material from which the respective RP parts are fabricated. In FDM the material is melted inside a feed chamber from which it is extruded. In SLS the material is fed in a powder form and a laser is used for selective melting of the layer cross-section. Both processes therefore result in parts that are relatively strong and heat resistant when compared with most of the other RP processes. Such parts are often referred to as having 'functional' properties that can be used for actual applications or at least for testing purposes.

Since FDM parts are produced using a material that is extruded onto a substrate platform, they require support structures for overhanging features in much the same way as SLA parts do. These supports must be removed from the actual part in the finishing operation, and there are currently two different techniques in use. One method uses a material with slightly different mechanical properties to form the support structures. This material breaks away quite easily from the part material when the part is completed. The second approach uses a material that is water soluble. After a short time in warm water, the part material remains. This process can be accelerated using ultrasonic agitation of the water. This second process leaves a cleaner interface region where it used to join with the part, and supports can also be removed from difficult-to-access regions within the parts. The first process, however, can still be useful,

Figure   1.2 Maxum machine, along with a spine model made using FDM (courtesy of Stratasys)

particularly for certain medical applications. Since the support material is extruded from a separate chamber, this material can be chosen to have a different colour. The model can be segmented in such a way as to require building using part material in some regions and support material in another. It is not essential that all the support material be removed, and so the segmented regions can be used to represent different features. For example, one material can be used to represent healthy bone while the other material can be used to represent cancerous material. While these materials are not transparent in the way SLA resins are, there is the advantage that each material can be chosen to have a colour from a range of choices. Since there are only two extrusion chambers in FDM, this is the maximum number of colours that can be feasibly chosen.

Since the SLS process produces parts from a powder substrate, selectively melting the material according to the corresponding cross-section, the major advantage is that supports do not have to be generated. The unmelted powder that surrounds the part is constrained within a build chamber which ultimately provides a natural support for the part during fabrication. This cleans away from the part very easily, reducing the post-processing time as well as the software set-up of the part (since no support data need be generated). Control of the temperature during the SLS process is very critical, and parts exhibit a powdery texture that is also slightly porous (for functional parts around 80%, which also roughly corresponds to the expected tensile strength of a part). The major advantage of this process is its versatility. Parts are generally more accurate than FDM parts, with strength properties acceptable for functional purposes (like cutting and screw fixation) when using a nylon-based material. The same process can also make parts suitable for investment casting applications using a styrene-based material. Change the material again and metal parts can be made with the addition of a furnace infiltration process. These parts are a blend of bronze and steel that is acceptable for short-run injection-moulding tooling applications. The investment casting process may be useful for making parts for medical applications since the metal can be chosen to be biocompatible.

Of course, with the added versatility of the SLS process comes added cost. The baseline FDM machine is considerably lower in price than the SLS machine. The other major difference between the two machines is the time to build. FDM machines are considerably slower than,

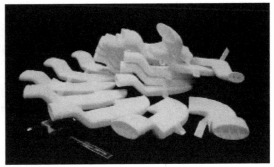

Figure 1.3 Selective laser sintering equipment and sample parts made from nylon powder (courtesy of 3D Systems)

in fact, most other RP processes. While SLS machines have a lengthy warm-up time, it is possible to build parts in batches, which means time per part can be quite low.

1.7 Droplet/Binder Systems

More recent systems have taken advantage of ink-jet or other forms of droplet deposition system to make RP parts. Droplet systems are very controllable, deposit materials in very small quantities and can be constructed in arrays and are therefore very suitable for forming the basis of a deposition system. There are two main ways in which droplet deposition is used: drop-on-powder and drop-on-drop.

Drop-on-powder systems use the ink-jet system to deposit a binder that glues powder particles together. The form of this technology developed for RP has its base in an MIT patent named 3D printing. MIT licensed this technology to a number of companies, the most widely known being ZCorp, which makes a range of high-speed, low-cost machines. Although parts from these machines exhibit relatively poor mechanical properties, they can be strengthened. The ink-jet binder printers can also deposit materials of different colours, thus making them useful for fabricating multicolour parts. These parts are opaque like the FDM process, but the multiple colours may benefit a number of applications where effective communication of ideas is required. Another licensee of the basic 3D printing process is Therics Inc. who have used the process to create controlled drug delivery devices and who are also investigating tissue engineering applications.

Drop-on-drop systems deposit the material in a liquid form, which then solidifies immediately after. This can be achieved with the base material in molten form, using cooling to solidify it, or using a photocuring process similar to SLA. In the latter process a strong curing light is used to solidify the material after it comes from the print nozzle. There are a number of machines using this approach to make parts. Since the machines use mass-produced printing technology, the cost can be kept quite low without compromise to accuracy. These machines are therefore generally low cost and fast acting (although there are some exceptions, particularly regarding speed of build).

Figure 1.4 ZCorp Z406 machine along with some parts made using the colour cabability (courtesy ZCorporation)

Parts that come off these machines are generally weaker than SLA, FDM or SLS parts. There is also a limitation in the range of materials available and therefore the corresponding applications. These machines are often referred to as 'concept modellers', their application being well suited to the early stages of product development. For medical applications one might see these machines being used as communication aids where a doctor can illustrate a proposed procedure to a patient with the help of a model of the patient data.

1.8   Related Technology: Microsystems and Direct Metal Systems

As already discussed, there are specific limitations for most RP technologies – those of accuracy and material properties. In general, most RP technologies work on the submillimetre scale, with most machines operating around 100 μm and the most accurate going down to around 25 μm resolution. Similarly, most RP models are made from polymer materials that have limited strength and thermal resistance when compared with metals, for instance. These limitations are a consequence of the layer-based approach and in particular how the layers are combined together. The high temperatures and forces that may be required to combine stronger materials together may also result in expensive technology and energy wastage. Similarly, if machines are built to fabricate parts in the low-micron or submicron scales, machines would require very expensive positioning and control.

Since many existing medical applications do not appear to require high accuracy at present, there are no chapters on this subject in this book. However, there may be some need for such technologies in the future, particularly where there may be a requirement for complex heterogeneous structures. When reading in this book about tissue engineering, I hope you will appreciate the complexity of tissue structures and that they have complex microstructures that may benefit from the developments in micro-RP. Therefore, it is important to note that this technology does exist and a number of systems are under development for manufacturing complex geometry using layer thicknesses of 5 μm or less.

There is a chapter in this book on direct metal fabrication. There is a definite and immediate need for the fabrication of custom artificial implants. While this may be achieved in the longer term by tissue engineering, this technology has many problems to overcome. Direct metal fabrication using RP techniques is also a technology that is still in its infancy, but there are already parts being made using a number of machines worldwide. The metals used can be chosen to suit different applications, and there are a number of metals that can be suitable for fabricating medical implants.

1.9 File Preparation

As we know, RP is a technology for converting 3D solid modelling, virtual, CAD files into physical models. Since there are many different types of CAD file format, it is necessary for CAD files to be converted so that the RP machine can read them. Nearly all RP machines use the same file input method for model files. This method is called the STL format after the stereolithography process that first defined it. STL files are surface approximations of the solid models generated within the CAD system. The approximation is performed by forming the closed surfaces using planar triangles. If the triangles are very small, then the surface approximation can be very accurate, although for curved and freeform surfaces (like medical models) the file sizes can be very large.

The reason for the triangle meshing is that it makes the slice preparation quite straightforward for the RP operating system. Slices are basically generated by checking the intersection of

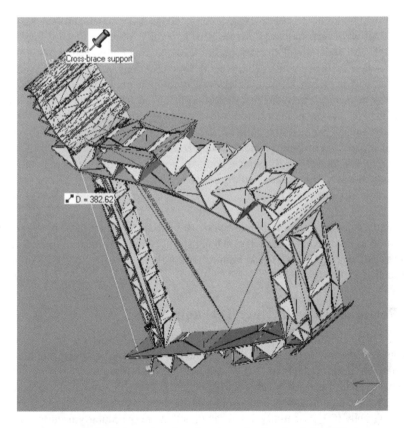

Figure 1.5 CAD part showing the triangular facets of the STL data

triangles with a plane that represents the Z height for a particular layer. Each intersecting triangle will form a vector that is part of an outline for the object layer. Vectors must be connected up to form a complete outline contour. Since it is possible there will be many contours forming a particular slice, it is also necessary to determine nested contours. Contours nested inside another represent a cavity or hole, for instance, while further nesting may represent other regions of the object to be fabricated. Conversion to STL format is a relatively simple process and there is generally no distinction between different objects. Separate objects could be combined into the same STL file. Sometimes this is helpful if the objects interlock in some way. Chain links, for example, could have each link physically separate from each other, but they can be interlocking to form a complete chain. The links do not require to touch each other in the STL file and so space can be formed around each element so that they can freely move within the chain. This could be helpful for complex medical modelling of anatomical joints or other interlocking bone regions.

The STL files, once read into the RP machine, generally require some software processing corresponding to the machine itself. Machines that can use different materials may also require different set-up parameters for each material, and these should be selected at this stage. Often there is a need to make an assessment of the complexity of the geometry to ensure the part is going to be built optimally. Some examples include:

- Orientation of build. Since machines are heterogeneous in character, parts can vary in quality if built in different orientations. Finer detail is generally found on upward-facing surfaces, for example. Also, parts will vary in build time, depending on the height of build.
- Thick and thin wall decisions. Thin-walled structures may require additional energy or material to ensure effective build. Thick-walled components may correspondingly require less energy or material to avoid part distortion or unnecessary wastage. In such cases, fill patterns may be chosen as an option.
- Complex geometry. Some RP machines may require consideration relating to the complexity of part geometry in terms of trapped volumes, overlapping features, etc.
- Support removal. Related to the above, complex geometry components may have supports that cannot be easily removed once the part is complete. Decisions on whether to modify or remove these supports must be made.

Many of these issues are particularly relevant to medical models since these are often complex, freeform models that are not designed using any specific engineering principles. Such models represent a challenge to RP model makers, and therefore considerable expertise is often required to build such models effectively.

1.10   Relationship with Other Technologies

It is important to mention at this stage that RP is only possible with the support of other technologies. In fact, it is necessary that the models be computer generated, and therefore the considerable developments in CAD/CAM are very significant to the development of RP. For medical models, this issue is particularly relevant because the source of model data can be manifold. Medical models are generally sourced from patient data that can come from computerized tomography (CT, essentially 3D X-ray data that are particularly used for bony tissue), magnetic resonance imaging (MRI, used to generate images of soft tissue regions), ultrasonic imaging, etc. There are variants of these methods, and indeed images can be combined from different sources. In addition, conventional engineering CAD models can also be combined with such data to produce designs for implants, tooling and fixtures.

One benefit of RP is that the technologies can be used at varying stages of product development. It is an aid for conceptualization, but can also be used to generate detail designs and test models, and even for production. Of course, not all RP machines can be used in the same way and some are less versatile than others. Also, it is important to realize they are not used in isolation, but in conjunction with other technologies, like vacuum casting, spray tooling and injection moulding. Considerable skill is also required to operate these technologies effectively, with a requirement for manual skills to produce good surface finish and appearance of models while maintaining accuracy.

1.11   Disadvantages with RP for Medical Applications

As was stated at the beginning of this chapter, RP was not developed to solve the problems of medical modelling; it is more happenstance that it is suitable for such. It therefore follows that RP is not an ideal solution and there are problems in using it.

The most obvious problem is cost. While it is difficult specifically to allocate cost to a process when considering the potential for saving or improving the quality of life, it is clear that all technologies have associated costs. Even where there are obvious benefits in terms of improving the medical service, the approach may be cost prohibitive. Many RP machines are costly to run, particularly in terms of material costs. This is partly a consequence of the relatively low number of machines currently available. As the technology becomes more popular in all areas, operating costs will surely drop. Indeed, evidence already shows that operating costs have dropped consistently year on year over the last 15 years.

In addition to cost, the properties of the materials used leave much to be desired. Most importantly, RP materials must be considerably more biocompatible than they are at present. Many materials are not even fit to be sterilized and taken into operating theatres. The mechanical properties of most RP parts are generally poor, with parts often being too weak or brittle to withstand constant use. One aim of RP research is to develop rapid manufacturing technologies that use the layer-based approach to direct manufacture of products. With the demanding environment that is associated with constant use, harsh and variable conditions and heavy physical loads, it is surely a long way off before rapid manufacture of medical devices is possible.

The term rapid prototyping is somewhat ambiguous. Sometimes the model is not used as a prototype (in the case of rapid manufacture) and, more importantly for many, the parts are not made rapidly enough. The term refers to an alternative way of producing prototypes which required significantly more time and effort from skilled artisans. However, this is something engineers and product designers and developers can appreciate more than doctors and surgeons. As it is, RP can only be applied to applications that involve planning over periods of weeks or months rather than emergency situations.

Finally, doctors and surgeons are supported by many different technical experts. It is difficult to see exactly which type of technician would be responsible for making models, but it is likely such expertise is not commonly available in a normal hospital. This may have an influence on the type of machine that would prove suitable for medical applications since the more versatile machines also require greater care, attention and expertise in order to run successfully.

1.12   Summary

This book contains a variety of chapters relating to the application of advanced manufacturing technology for medical applications. The first few chapters are from practitioners using RP in their work with patients, explaining the purpose of various tools as well as generally providing us with the benefit of their experience. These authors are true pioneers in this field and we are extremely lucky that they agreed to collaborate on this book.

It is probably true to say that none of the authors is very far from the applications, even those who are obviously describing ground-breaking research. It is vital these techniques they are developing are relevant to medical procedures. Otherwise, no matter how impressive they look, no surgeon would use them. All the researchers are therefore working very closely with medical practitioners to ensure their work is both convincing and beneficial. Later chapters are therefore still very careful to point out the benefits to the user.

This book does not just focus on RP. There are significant sections on the imaging or reverse engineering tools used to prepare and process the data as well as discussions on the post-processing requirements for medical models.

I hope you benefit from this book. If you have a medical background, I hope it inspires you to consider using these new technologies to provide better care for your patients. If you are an engineer or a researcher, I hope it provides you with an insight into the state of the art and guides you in your future research.

Bibliography

The following books, journals and websites are recommended if you wish to learn more about rapid prototyping.

3D Systems, *www.3dsystems.com*

Burns, M. (1993) *Automated Fabrication: Improving Productivity in Manufacturing*, Prentice Hall, ISBN 0131194623, 369 pp.

Campbell, R. I. and Gibson, I. (eds), *Rapid Prototyping Journal*, Emerald, ISSN 1355 2546.

Chua, C. K., Leong, K. F. and Lim, C. S. (2003) *Rapid Prototyping – Principles and Applications*, World Scientific, ISBN 981 238 120 1, 420 pp.

Gibson, I. (ed.) (2002) *Software Solutions for Rapid Prototyping*, Professional Engineering Publishing, London, ISBN 1 86058 360 1, 380 pp.

Hilton, P. D. and Jacobs, P. F. (2000) *Rapid Tooling: Technologies and Industrial Applications*, Marcel Dekker, ISBN 0824787889, 270 pp.

Jacobs, P. F. and Reid, D. T. (1992) *Rapid Prototyping & Manufacturing: Fundamentals of Stereolithography*, Soc. Manuf. Engineers, ISBN 0070324336, 434 pp.

Lü, L., Fuh, J. Y. H., Wong, Y. S. (2001) *Laser-induced Materials and Processes for Rapid Prototyping*, Kluwer Academic Publishers, ISBN 0792374002, 267 pp.

McDonald, J. A., Ryall, C. J. and Wimpenny, D. I. (2001) *Rapid Prototyping Casebook*, Professional Engineering Publishing, London, ISBN 1 86058 076 9, 260 pp.

Stratasys, *www.stratasys.com*

Wohlers, T. T. (2004) *Wohlers Report – Rapid Prototyping, Tooling & Manufacturing State of the Industry Annual Worldwide Progress Report*, Wohlers Associates, Inc., 113 pp.

Wohlers Associates, *www.wohlersassociates.com*

ZCorporation, *www.zcorp.com*

2

Role of Rapid Digital Manufacture in Planning and Implementation of Complex Medical Treatments

Andrew M. Christensen and Stephen M. Humphries

The evolution of medical imaging has provided clinicians with the means for detailed views of the structure and function of patient anatomy. Further, software packages that construct complex virtual models using this image data are in common usage not only for qualitative visual inspection but also for quantitative planning of treatment parameters. While the combination of image data and computer modeling yields a variety of tools for visualization and treatment, difficulties remain in designing and implementing preplanned treatment parameters.

Advances in manufacturing technology have made possible the fabrication of highly accurate, custom physical models of patient anatomy and patient-specific treatment aides. For example, rapid prototyping (RP) generated anatomical models translate image data into solid replicas giving surgeons the means for tactile interaction with patient anatomy prior to the operation. Such models are used frequently to plan complex craniofacial surgeries, and as the basis for customization of devices such as titanium plates. In addition, treatment devices that incorporate patient-specific anatomical features and preplanned treatment parameters such as drill trajectories or osteotomy planes give surgeons elegant and reliable tools to carry treatment designs from the computer to the operating room (OR).

Advanced Manufacturing Technology for Medical Applications   Edited by I. Gibson
© 2006 John Wiley & Sons, Ltd.

2.1 Introduction

Physicians engaged in complex medical interventions such as cranial reconstruction, orthognathic surgery and radiotherapy, among others, increasingly rely on technologies that facilitate preprocedural planning and simulation. Parallel progress in imaging and computer techniques for virtual data manipulation provides the framework for detailed planning of complex medical treatments. Despite the increasing levels of detail possible with more and more powerful computer hardware and software, virtual representation of medical image data remains, for the most part, limited to on-screen displays that provide no means for true 3D understanding or tactile interaction. In addition, the transfer of treatment parameters from the planning environment to the actual patient for intervention is challenging. Virtual planning environments often allow data visualization and manipulation on a scale that is much finer that can be appreciated practically. In addition, most virtual planning systems are unencumbered by such real-world challenges as patient motion, bleeding, limited visibility through small incisions and sterility issues. Therefore, while it may be possible, for instance, to plan the perfect trajectory for a spinal pedicle screw in a virtual environment, actually implementing this treatment plan in a stable, repeatable and reliable fashion is more difficult. There is certainly a value in the exercise of virtual planning, in that the surgeon will have the opportunity to gain understanding of unique patient anatomy. This can be a valuable preparation tool for physicians, providing a detailed preview that complements their trained expertise. As the trend towards less invasive and time-consuming procedures continues, physicians will likely rely more heavily on advanced tools, including use of digital manufacturing, for planning and delivery treatment interventions.

Advanced manufacturing methods including rapid prototyping (RP) afford the means to fabricate solid objects based on medical imaging data. RP-generated anatomical models provide the means for tactile interaction with anatomy, and for rehearsal of tasks like osteotomy and surgical planning utilizing the actual instruments that will be used in a procedure. Models are used frequently in fields such as cranio-maxillofacial surgery, giving surgeons a highly accurate replica of patient anatomy for surgical rehearsal and customization of treatment devices.

RP techniques can also be used to create custom treatment devices that incorporate anatomic features, in addition to designed treatment parameters such as drill trajectories, cut planes and movement of bone segments. These rapid digital manufacture (RDM) techniques present elegant solutions to many medical problems, including the design of patient-specific implants and robust vehicles to bring treatment parameters to the patient in a truly custom way. Fabrication, through RP or RDM, of custom treatment aides such as templates, drill guides and cutting jigs provides specialized intermediary devices for accurate implementation of the custom-designed treatment intervention.

This chapter gives a general overview of the current status of computerized planning for medical intervention, with an emphasis on the role of RDM.

2.2 Primer on Medical Imaging

During the last weeks of 1895, Wilhelm Roentgen performed numerous experiments to characterize the new kind of rays he had serendipitously discovered. One of the most striking results of these experiments, an image of the bones in his wife's hand, demonstrated how the properties of X-rays could be harnessed to create pictures of internal bony anatomy. A new era in medicine had arrived.

X-ray imaging, or radiography, quickly became an important tool for physicians. During the decades after its discovery, developments such as contrast agents enabled the imaging of structures besides the skeletal system, broadening the scope of radiography. More recent advances such as fluoroscopy and computed radiography (CR) enable digital processing of radiographs. However, the inherent 2D projection geometry of radiographic imaging remains an obstacle to the use of these images for more quantitative planning. Stereotactic techniques, whereby reference markers are rigidly fixed to the patient's bony anatomy, can provide a basis for calculation of 3D positions from an orthogonal pair of radiographs, but this process is cumbersome, requiring detailed geometric calculations, and does not provide complete 3D information.

Since its development in the early 1970s, computed tomography (CT) has evolved into a comparatively inexpensive and fast tool for acquisition of detailed scans of patient anatomy. Current multislice scanners are able to scan all or large portions of anatomy quickly and at high resolution. Fast scans with very thin slices, which return 3D volumes of image data, are opening up applications in CT angiography, cardiac scanning and trauma. CT image data are relied upon extensively for many medical and surgical purposes, including diagnosis (i.e. visual inspection of image slices), and for the creation of virtual models for planning. Examples include stereotactic and image-guided surgery, radiotherapy and calculation of tumor volumes. In addition, quantitative CT (qCT) is increasingly popular for calculation of bone density. In qCT, Hounsfield numbers are calibrated to mineral densities by scanning tubes with known concentrations of minerals in solution. The qCT technique is relevant to this discussion because it extends this imaging modality beyond the simple generation of images of internal anatomy into the realm of non-invasive testing of physiology.

Cone-beam CT (CBCT) is a technique for 3D X-ray imaging whereby a large X-ray field, emitted in a cone shape, rotates about a subject in conjunction with an opposed 2D detector, such as an image intensifier (II) or amorphous silicon detector. CBCT enables acquisition of a full 3D volume of data in a single rotation of the X-ray source and detector, compared with the slice-by-slice acquisitions in traditional CT (or the slab-by-slab acquisitions of multislice CT). CBCT requires significant computing power to process large volumes of data but has the potential to acquire significant amounts of that data rapidly while exposing subjects to a smaller dose of ionizing radiation than traditional CT (Danforth, Dus and Mah, 2003). Currently, CBCT devices are commercially available for head and neck and intraoperative imaging at very high spatial resolution, but with lower contrast resolution than traditional CT. This can be expected to improve with refinements in detector technology.

Invented in the 1980s, magnetic resonance (MR) imaging employs strong magnetic fields and radio waves to produce detailed images. MR imaging utilizes the behavior of hydrogen protons in water molecules in the body and tends to reveal more soft tissue information than CT, which measures X-ray attenuation and therefore only shows variations in tissue density. MR imaging can reveal significant soft tissue detail such as ligaments and tendons as well as the presence of tumors. Image acquisition parameters can be tuned in many different ways, so it is possible to acquire a variety of different image data while the patient is in the scanner. Examples include proton density scans and MR angiograms. A more recent variant of MR is functional MR (fMR) which detects subtle changes in magnetic fields at a molecular level that can indicate greater blood flow to certain areas of the brain. The results of fMR can indicate which areas of the brain are active while a subject is performing a particular task. This information can be quite valuable in evaluating patients with stroke or brain tumor, and its use in planning surgery and radiation therapy is increasing. Similarly, MR spectroscopy is developing into an imaging modality that can provide indications of the concentrations of different chemicals revealing pathological processes in the body non-invasively.

Positron emission tomography (PET) has become increasingly popular and is frequently used to measure functional and metabolic activity. In PET imaging a radioactive agent is tagged to a material that is biologically processed by the patient. Radioactive oxygen-15, which a patient inhales prior to a functional brain scan, would be one example. When the subject performs a particular task, like speaking, the oxygen isotope is delivered to the areas of the brain involved in that task, and its radioactive emissions are measured and reconstructed in a set of tomographic images. PET scans are also used to stage lung and liver tumors, meaning that they show which parts of a tumor are most metabolically active and the extent of the cancer. This information is used increasingly to facilitate treatment planning, so interventions can be directed specifically at the most active portions of a tumor.

Ultrasound (US) imaging plays an important role in several medical specialties, largely because it does not use harmful ionizing radiation and can be performed quickly under a variety of circumstances (i.e. in office, in OR). Newer US devices can image in 3D and in 4D, meaning that 3D images are acquired rapidly over time, providing a 3D cine view. With calibration, US image data can be correlated with CT or other imaging modalities to be used quantitatively for treatment planning and verification.

Many types of non-medical device have been reverse engineered for use in the medical sector. Scanning technology that records surface topology optically, using lasers, or physically, using articulating arms, provides a means of gathering significant amounts of geometrical information without exposing the patient to the ionizing radiation of CT, X-ray and PET. Another advantage is that image data of relevant anatomy may not be subject to digital artifacts from, for instance, metallic implants in teeth which can cause streaking in CT. Also, 3D color scanners can gather surface color data to create realistic virtual models of patients' faces. Several groups are investigating the use of this type of device in cranio-maxillofacial (CMF) surgery (Yamada *et al.*, 2002).

2.3   Surgical Planning

2.3.1   *Virtual planning*

Historically, and, to some extent, currently, physicians have relied upon training and expertise, sketches, simple models and mental visualization to plan for their procedures. While this approach remains, to a degree, the standard of care today, experience indicates that the bar will be raised once surgeons fully appreciate the power of the computer and imaging modalities described previously in planning and executing surgical interventions. As the trend towards less invasive, more precise treatments continues, advanced technologies, capable of integrating all stages of a procedure (planning, delivery, follow-up), will assume central importance.

Software systems exist that can load, integrate and manipulate different types of image data to construct complex virtual models. By stacking co-registered tomographic image slices into a single volume of data, 3D renderings and oblique slice reconstructions become possible (Figure 2.1). Radiologists often evaluate such displays in conjunction with the original image slices for diagnostic purposes. Most modern CT and MR scanners are equipped with such software at the operator's console, so that secondary reconstructions can be calculated from the

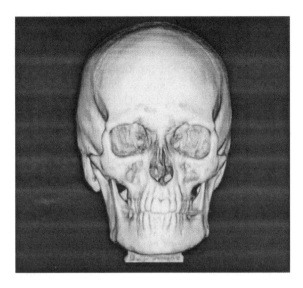

Figure   2.1 Computer-generated 3D rendering of bone, based on CT scan images

primary image data at the time of their acquisition. Qualitative evaluation of these reconstructed images and volume renderings increasingly plays a role in the radiologist's practice. Certainly, the ability to acquire and process images at higher and higher spatial resolution has pushed the development of new technologies and applications (Gateno, Teichgraeber and Xia, 2003a). Physicians have been forced to follow suit.

There are noteworthy limitations to the current status of 3D image reconstructions and display, however. In particular, the systems do not provide tactile interaction with the data, and renderings on computer screens do not show true 3D relationships, instead utilizing computational models of light and shade to render images on a flat monitor that the observer perceives as 3D. While technologies such as haptic interfaces, whose motors and sensors provide the basis for mechanical feedback, and volumetric displays are emerging, they are not in widespread use.

One of the earliest applications of computer-aided quantitative treatment planning in medicine was arguably the design of radiation therapy dose plans (Worthley and Cooper, 1967). Computers were applied to the task of predicting, through complex calculations, radiation doses deposited in tissue when exposed to combinations of radiation beams. Radiation oncologists and physicists would rely on these computational results to design beam arrangements that delivered a prescribed dose of radiation to a target volume while avoiding overdose in the surrounding healthy tissue. The earliest versions of such programs calculated doses at single points and relative only to coarse representations of patient anatomy such as a single CT slice or a digitized patient contour (acquired with a plaster of Paris strip). Current radiation therapy treatment planning systems calculate full 3D distributions of radiation doses and can generate detailed renderings built from co-registered sets of multimodal image data. What is noteworthy in this discussion about radiation therapy treatment planning is that image data are used to construct a virtual model of a specific patient and complex algorithms are used to compute results (radiation dose distributions) of a given treatment option. Typically, several treatment designs are simulated and the plan that best fits the physician's prescription is implemented.

Image-guided surgery (IGS) systems were developed in the early 1990s as a technique to link image data and virtual models with actual patient anatomy (Smith, Frank and Bucholz, 1994). IGS systems incorporate spatial tracking systems, typically optical or magnetic, with software systems that handle medical image data. The result can be considered analogous to a global positioning system (GPS) for the operating room (OR). In GPS, satellites orbiting the Earth can localize the position of a GPS transmitter in a car, correlate its position to a street map and provide directions to the driver. In IGS, localizer technology can track the location of an instrument in a surgeon's hand, correlate its position relative to preoperatively acquired images and provide guidance to the surgeon. IGS systems are used by neurosurgeons to localize brain tumors for less invasive and more complete resections, by ENT surgeons for safer, more precise sinus operations and increasingly in orthopaedics for guidance in total joint replacements.

By necessity, IGS (so-called 'surgical navigation') systems apply advanced software features to the reconstruction and rendering of complex image datasets. These features permit quantitative planning of surgical trajectories and the measurement of target volumes but, in fact, are typically used in only a few types of procedure. In cases such as stereotactic biopsy, or implantation of deep brain stimulators, planning tools are used to determine settings for traditional stereotactic devices. In most other types of case, such as functional endoscopic sinus surgery, which is a much faster procedure, performed using free-hand instruments, little or no computer planning is performed before the procedure.

2.3.2   Implementation of the plan

Just as the field of radiation therapy is a good example of early computer treatment planning, it is also an appropriate example of the challenges of implementing preplanned treatment parameters (Miralbell *et al.*, 2003). Even in modern radiation oncology departments there is a heavy reliance on traditional manual techniques to create custom patient support and alignment devices designed to enable delivery of treatments designed virtually. Moldable thermoplastic materials, pillows and multiple sessions for simulating treatment set-up and delivery are necessary actually to implement a treatment plan generated in the computer. This reliance on manual techniques to create devices to facilitate implementation of a treatment plan leads to a 'digital disconnect'. Treatments are planned in a virtual environment and delivered using complex computer-controlled devices. However, the intermediate steps to deliver treatment rely on manual transfer of information and fabrication of devices, which obviously can be subjective and prone to human error.

Many types of rigid but adjustable device have been developed to facilitate precise delivery of treatment parameters that were designed prior to the procedure. Stereotactic frames are a good example of this (Figure 2.2). Used for targeting in radiation treatment or surgical intervention, stereotactic frames are rigidly fixed to a patient's head using invasive pins. Imaging studies acquired with the frame in place include the patient's anatomy as well as the stereotactic frame, so it is possible to integrate target point coordinates from the image system into the coordinate system of the frame. With the position of the surgical target established relative to the position of the externally fixed frame, it is possible to adjust the settings of the stereotactic aiming device to reach these targets. The drawbacks to such techniques are that imaging studies must be acquired with the frame attached, which is cumbersome and uncomfortable, and that only

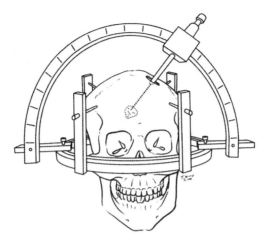

Figure 2.2 Stereotactic frame for surgery

one point or trajectory can be targeted at a time. Also, because they are invasive, it is not possible to use such frames in fractionated treatments. A myriad of devices for the non-invasive, repeat fixation of stereotactic implements have been proposed and patented (Sweeney *et al.*, 2001). Most use moldable bite blocks, plaster or fiberglass casts, straps and/or ear plugs to stabilize the patient's head during imaging and treatment. These tend to be very dependent on the skill of the user, rely on patient compliance and are less precise than true stereotactic frames.

Surgical navigation systems are thought to be an advance over stereotactic frames, since they avoid bulky invasive devices and can track instruments in real time. However, they do have some significant weaknesses, chief among which is the free-hand nature of the instruments. The surgeon must manually align an instrument to the preplanned trajectory, using on-screen displays for reference. This can be difficult and draws the surgeon's attention to the computer screen and away from the actual patient. Even when the surgeon aligns the instrument to preplanned parameters, this position is lost when the surgeon sets the instrument down. The practicalities of surgery require that surgeons often switch instruments and that they maintain attention on the patient directly, rather than to a virtual representation of the patient. For example, in most types of surgery, tasks like the management of bleeding are not guided by presurgical planning and surgeons must focus on the patient's anatomy as it appears in front of them, rather than on a computer screen or some other representation of the patient. This causes an 'attention split' problem, as surgeons must divide their attention between the virtual patient and the actual patient. While using the computer monitor to implement the plan and align the instrument, the surgeon is still responsible for performing the basic surgical management tasks at hand.

There are a few developing technologies that have been applied to the task of implementing a preoperative treatment plan. Robotics would seem to be a means of compensating for some of the limitations mentioned, but for the most part it has not been embraced by the medical community (Honl *et al.*, 2003). The ability to drive a trajectory guide or instrument carrier into a position defined by a preprocedural plan would avoid the limits of free-hand navigation (Choi, Green and Levi, 2000). A robotic arm could be repeatedly moved away from the surgical field, then replaced, thus alleviating the difficulties of working with, for example, a stereotactic frame. Despite its potential advantages, robotic technology has not gained much acceptance, likely because of high costs, limited applicability (devices that have been marketed tend to be suitable for only a small number of procedures) and concerns over reliability and the loss of human control.

Another possible solution to the attention split problem is augmented reality, that is, the ability to project computer-generated data into a surgeon's field of view in real time. This is a developing technology used in defense and industry that would provide guidance by overlaying graphic representations of virtual preplanning directly into the surgeon's field of vision. In other words, preplanned parameters would appear to the surgeon directly in the field, rather than on a screen. However, no physical guidance is provided and instruments must still be manipulated free-hand, imposing limitations similar to some of those referenced above.

2.4   RDM in Medicine

2.4.1   RP-generated anatomical models

With the advent of computed tomography scanning techniques in the 1970s, surgeons were given the first glimpse at spatially correct datasets produced non-invasively. This revolutionized the diagnosis and treatment of many different pathologies, and its scope has continued to expand since that time. Very early on, some surgeons stacked the bone structure image data, slice upon slice, to create crude 3D images. This allowed for unprecedented visualization in 3D and made possible the evaluation of pediatric deformities and traumatic injuries in an entirely new way. This modality was extremely useful, but still lacked the tactile element a physical model would provide.

Since the early 1980s there have been several methods for creating 3D physical models from imaging modalities such as CT scanning. Companies have produced these models for several years by milling materials such as foam and polyurethane. Milling is a subtractive process in which the model is formed by the controlled removal of material from a block of that material. While this technique can ultimately provide a fairly precise representation of a 3D structure, it lacks the ability to provide accurate internal and external structures for the assessment of the pathology presented.

Stereolithography (SLA), the first so-called 'rapid prototyping' process, was developed in the late 1980s to mitigate the weaknesses of the milling process. SLA allows designers to 'print' in 3D the parts they design. The process employs ultraviolet lasers which initiate a photopolymerization reaction that locally cures liquid resin into a solid plastic (Figure 2.3). Most of its immediate use was in the automotive and aerospace sectors, but gradually medical applications of this technology emerged. By the time 3D CT had been in use for about 10 years, surgeons in certain specialties were relying more heavily on 3D visualization tools. The merging of 3D medical imaging and stereolithography was fairly easy because both technologies relied on slices, or layers, as input. Software tools were developed specifically for this task, and surgeons have come to rely on this technology as a standard element in the planning of some complex treatments.

Stereolithographic anatomical modeling has been used in clinical practice in some form since the early 1990s. In basic terms, the process uses computed tomography (CT) images and computer-assisted machinery to produce physical models of the bone structure of a particular

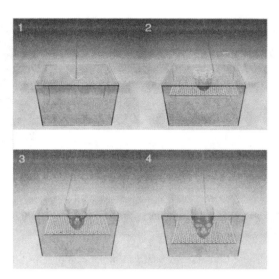

Figure   2.3 Schematic of the stereolithography process, shown producing a skull

patient. Surgeons have found uses for these tactile models in many specialties, but cranio-maxillofacial surgeons remain the largest group of their users worldwide.

This technique provides models that are extremely accurate and allow for visualization of internal structures, such as nerve canals and sinus cavities. Figure 2.4 shows an example of a stereolithography model produced from CT scan data. Rapid prototyping (RP) is characterized

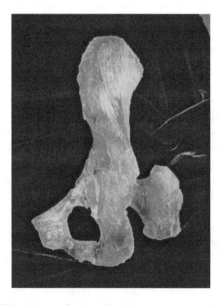

Figure   2.4 SLA model of the bone of a hemipelvis, produced using CT images (courtesy of Medical Modeling LLC)

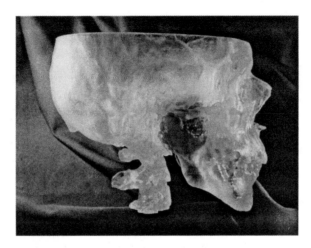

Figure 2.5 3DP model based on skin surface and bone as shown in CT (courtesy of Medical Modeling LLC)

by processes that create a physical object from computerized data using an additive, layer-based technique. The technology has evolved to encompass a family of manufacturing techniques including selective laser sintering (SLS), fused deposition modeling (FDM), 3D printing (3DP), laminated object manufacturing (LOM), multiJet modeling (MJM) and many others. Specialized software interfaces are required to take the medical imaging data typically used for anatomical modeling (i.e. CT or MRI) and convert it into the files needed to guide the RP apparatus.

The models produced are typically of the hard-tissue anatomy because of the need to reconstruct osseous defects. The appropriate imaging modality for these models is CT or CAT scanning, which, interestingly, can also be used to create soft-tissue models, most specifically of the external anatomy. Figure 2.5 shows a produced model of an infant with a severe craniofacial disorder. Occasionally, vasculature structures need to be modeled. A CT scan with a contrast agent is then used to locate the tumor, for its segmentation by the computer software. Figure 2.6 shows a model of a patient with a myxoma involving the right temporomandibular joint and cranial fossa. Using a two-color modeling process that has been developed, it is possible to segment separate structures apart from the bone. This is made possible by a specially formulated liquid resin that has two dose response levels to UV radiation. The first causes it to solidify and, if exposed to an additional dose from a UV laser, it will change color. In this case, the tumor shows up as a red material. This unique approach can also be used effectively for identification of the inferior alveolar nerve structures, tooth structures and soft- and hard-tissue tumors as well as existing implants.

Physical anatomical models (biomodels) produced from medical imaging data have been used more and more frequently over the last few years. These models are typically of the bony anatomy and are used for planning complex reconstruction cases which may or may not involve customization of a surgical device or implement to be used during the treatment. These models provide something that no other imaging study can: a physical object from which to make measurements; a tactile replica about which to bend devices; a simulacrum upon which to rehearse procedures, using actual surgical instruments (Christensen *et al.*, 2004.)

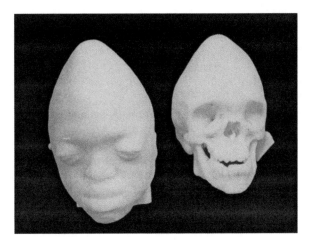

Figure **2.6** Two-color stereolithography model showing a myxoma tumor of the right temporo-mandifular Joint (courtesy of MedicalModeling LLC)

Early uses for physical modeling included the design and fabrication of custom-made titanium mesh for cranial defects and mandibular trays. Models have since been used for the design and fabrication of implants ranging from total temporomandibular joint (TMJ) replacements to partial knee replacement devices. Apart from the creation of truly custom implants, models have been used increasingly for procedures such as jaw surgery or spinal fusion in planning the procedure.

Distraction osteogenesis of the facial skeleton can require more complex planning than traditional orthognathic surgery because of the gradual lengthening process and the devices to be manipulated. The learning curve for distraction is reportedly steep, resulting in a higher rate of complication when inexperienced surgeons perform the procedures than when more experienced surgeons perform them. Stereolithographic models have been used when planning distraction procedures in order to moderate the learning curve. A CT-based physical model allows for true 3D visualization, osteotomy planning and device adaptation to the anatomy prior to the procedure. This has been extremely helpful in saving time during Le Fort III or monobloc advancement cases utilizing distraction. Selectively colored stereolithography models highlighting the inferior alveolar nerve, teeth and teeth buds have also been used for precise planning of distractor attachment to avoid damage to vital structures.

Companies that produce and supply RDM models are called service bureaus. A competent service bureau will employ a skilled radiological staff, trained to image the human body and empowered to make informed decisions about the production of these highly accurate models. Because the machinery is very expensive to purchase and maintain, most hospitals and medical centers are currently unable to bring this technology in-house. The first step in the process of acquiring a model involves the transfer of the data from the CT scan to the modeling facility. There is typically a lead time of 1–5 days for the production of a model from raw image data. Increasingly, these data can be transferred over the internet with file transfer protocol (FTP) directly from the radiology center. Reimbursement for models varies by country, with some health insurance providers covering the cost and some not covering the cost of models. Typical prices for anatomical models range from US$500 to US$3000.

Models are now available from medical modeling vendors in different materials. New processes such as 3D printing have allowed for less expensive models to become available for less complex cases. Models are also available that allow for intraoperative use. This expanded capability encourages surgeons to use models innovatively. One such use has reduced the morbidity of tissue grafts and their recipient sites by allowing for the manipulation of the grafted tissue upon a sterilized model.

The CT scan is the most important step of the entire modeling process. The ultimate accuracy of the model is utterly dependent upon the quality of the CT scan. The precise implementation of the modeling protocol is the second determinative factor, including the use of 1 mm × 1 mm continuous axial slices and a standard algorithm for the craniofacial skeleton. Other areas of anatomy, such as the pelvic bones, can be successfully imaged using 3 mm × 3 mm continuous axial slices which are subsequently reconstructed to less than 3 mm spacing. A competent service bureau providing modeling services should be able to provide specific protocols for many standard areas of anatomy prior to scanning.

2.4.2   Custom treatment devices with ADM

Advanced manufacturing techniques can also help 'close the loop' between digital planning and treatment delivery. There are several examples of RP technology used for fabrication of devices to assist in the implementation of surgical interventions. Drill guides for dental implants (Figure 2.7), stents used in CMF surgery (Gateno *et al.*, 2003b) and jigs to help align bone segments are a few examples. Such treatment aides incorporate patient-specific anatomical features, typical negatives of unique bone structure that are complementary to anatomy being targeted in the procedure. For example, drill guides used for dental implants fit snugly over bone of the jaw in only one possible position. Once in position the holes in the guide, which correspond to planned implant trajectories, constrain the physician's drill along the paths that were designed before the procedure in the computer. With the drill guides in place, the physician can quickly and confidently drill into bone without using traditional alignment techniques or

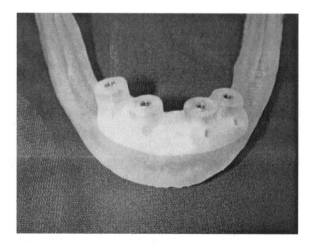

Figure   2.7 Surgical drill guide for dental implant placement

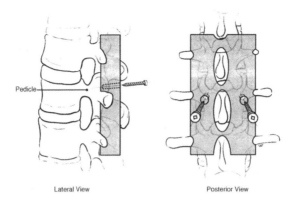

Lateral View   Posterior View

Figure 2.8 Concept sketch of template for pedicle screw insertion

online X-ray imaging. This is an excellent example of how modern manufacturing can facilitate the implementation of a treatment plan that was designed virtually. Similar types of custom device have been suggested for spinal pedicle screw placement (Figure 2.8) and surgery for pelvic fractures. Along the same lines, templates for volumetric resections in neurosurgery have been proposed. These devices match the contours of surrounding bony anatomy and include a window cut out to indicate the borders of a lesion to be resected.

Custom stereotactic devices manufactured using RDM techniques have been suggested and are commercially available (D'Urso *et al.*, 1999; Swain, 2004). These often rely on screw-in fiducial markers placed in the patient's head, surrounding the proposed surgical entry site, prior to imaging. These scans are then used as the basis for planning the trajectory of a biopsy needle or stylus which will introduce an electrical stimulator to a deep nucleus of the brain. Once this trajectory is planned with respect to the image data, and therefore the fiducial markers are also present in the scan, a physical guide can be manufactured (FHC Crop, Bowdoinham, ME, United States). This guide will mount to the fiducial markers, still present on the patient's head, and serve to aim instruments at the planned target point intraoperatively. The benefit of such an approach is that a bulky, invasive frame is avoided and the patient need not remain in the hospital between imaging and actual treatment.

The use of technologies generically grouped in the area of 'rapid digital manufacture' is growing in many sectors of industry, changing the way people buy products and services. The ability to custom-manufacture goods quickly and cost-effectively has aided several mainstream product lines and has impacted the production of consumer goods such as computers and running shoes. In the past, while the processes were too slow and expensive to allow their efficient, profitable use in consumer applications, they did make inroads into the clinical realm. One of the original uses of physical modeling from medical image data was the production of custom alloplastic implants and surgical implements. Notable early applications included custom total joint replacements like knee and hip prostheses as well as patient-specific TMJ replacements.

One of the most successful applications of RDM is the orthodontic product Invisalign (Align Technologies, Sunnyvale, CA, United States). In this application, 15–20 sets of tooth aligners serve the same purpose as traditional orthodontic braces. The process begins with a mold of the patient's teeth which is digitized to provide a virtual replica for planning the steps of straightening the teeth. From the computer plan a series of aligners are designed, which the patient wears in sequence in a process that lasts as long as 18 months. The aligners are not directly produced using stereolithography, but master models on which the designs are manufactured are created using stereolithography. The aligners are removable and translucent which provides many benefits in terms of comfort and cosmetic concerns.

Other kinds of computer-controlled manufacturing technique are used in medicine for customization on a large scale as well. Complex radiation treatment plans, specifically those employing a technique known as intensity-modulated radiation therapy (IMRT), are often implemented with custom-milled tissue compensating filters. With their design based on CT data and virtual treatment planning, these filters shape radiation beams in order to deliver treatment plans that seek to maximize dosage to tumor volumes while generally sparing the surrounding healthy tissue. The filters can be milled from metals that attenuate radiation beams (SPR, Sanford, FL, United States) and are thus able to vary the intensity of radiation across a beam, and to define the aperture of that beam (Zelinski, 2003). Computer treatment planning systems are increasingly automated, utilizing optimization algorithms to determine an arrangement of beams that will fulfill a physician's prescription. These systems provide clinicians with a virtual environment in which to design complex arrangements of beams. Often, equally complex devices are required to implement these virtual plans, and ADM techniques such as computer numerical control (CNC) machining is used with success. After a computer plan is approved by a physician, geometrical calculations convert the treatment plan to CNC instructions which can then be transferred electronically to a machining facility.

2.5   The Future

As technology inevitably charges forward, applications of RDM in medicine will expand. Research in biomaterials science indicates the feasibility of direct fabrication of implantable materials using RP-like technology. This could be an important advance allowing rapid manufacture of truly custom implantable devices such as joint prostheses, complete sections of missing bone or absorbable scaffolds that provide structure while engineered tissue replacements take hold. Instead of alloplastic materials, which are merely tolerated by the body, these newer materials have the potential to stimulate growth and actually be incorporated into the body.

RP-generated models and treatment aides are for the most part manufactured by specialized service bureaus. Even with modern techniques for rapid data transfer via the internet and reliable overnight shipping there inevitably is a lead time of several days for a service bureaus to provide a model. As more specialties capitalize on rapid manufacturing, it may be cost effective for health care facilities to arrange for an on-site or nearby RDM shop. It is expected that, if the amount of time required to get a model or treatment device in the hands of a physician is reduced, they may find a wider acceptance, particularly in fields constrained by time urgency such as orthopedic trauma or cardiothoracic surgery (Brown *et al.*, 2003; Brown, Milner and Firoozbakhsh, 2002).

Computer power has increased exponentially over the last decade. Relatively inexpensive computers are now capable of handling the large amounts of data produced in medical imaging. Off-the-shelf graphics capabilities now produce fast renderings from image data and allow the construction of composite models built from multiple image datasets. Simulation of treatment processes (calculation of radiation dose distributions, movement of bone segments and simulation of skin/soft-tissue response, biomechanical calculation of joint function, finite element analysis of strength, etc.) allows physicians to simulate the effects of different combinations of treatment parameters before deciding on the best course for the patient (Xia *et al.*, 2001). Also noteworthy are advances in applied mathematics and operations research, which have facilitated the development of optimization methods that can be applied to medicine. Optimization algorithms such as simulated annealing and linear programming can be used to find the best combination of parameters to produce the desired outcome. An important example is so-called 'inverse planning'. This technique, used in radiotherapy, makes it possible for a physician to input the desired dose to a target volume and the constraints, such as maximum tolerable dose to surrounding healthy anatomy. The optimization algorithm will then return a combination of radiation beams that would most closely achieve those results. In the traditional 'forward planning' approach, dose distributions resulting from different combinations of beam angles and intensities are compared by a human operator on a trial and error basis, until an acceptable plan is reached. Optimization algorithms will likely become more widely used in other medical specialties as well, allowing medical specialists to plan unique, mathematically optimal treatment strategies, rather than the field-fit approaches often arrived at by human trial and error. The results of such computer-generated treatment plans may very well be difficult to implement using traditional techniques. As such, RDM techniques could play a role in integrating the steps of such high-tech, computer-designed medical interventions.

2.6 Conclusion

Imaging and computer power have provided foundations for computerized data visualization and tools for preprocedural planning. While these are very useful on their own, they do not provide physicians with tactile interaction or the tools for true surgical rehearsal and 3D understanding. In addition, implementing treatment parameters that have been designed in a virtual environment is often difficult. ADM techniques have developed rapidly and provide means for fabrication of solid replicas of anatomy as revealed by medical image data. In addition there are several examples of the successful use of modern manufacturing techniques to facilitate the delivery of complex treatment plans. These techniques can be very helpful in bridging the gaps in current digital planning and treatment delivery techniques and promise to have an increasing role in medicine in the decades to come.

References

Brown, G. A., Firoozbakhsh, K., DeCoster, T. A., Reyna Jr, J. R. and Moneim, M. (2003) Rapid prototyping: the future of trauma surgery? *J. Bone Joint Surg. Am.*, **85A**, Suppl. 4, 49–55.

Brown, G. A., Milner, B. and Firoozbakhsh, K. (2002) Application of computer-generated stereolithography and interpositioning template in acetabular fractures: a report of eight cases. *J. Orthop. Trauma*, May, **16** (5), 347–52.

Choi, W. W., Green, B. A. and Levi, A. D. (2000) Computer-assisted fluoroscopic targeting system for pedicle screw insertion. *Neurosurgery*, October, **47** (4), 872–8.

Christensen, A. M., Humphries, S. M., Goh, K. Y. and Swift, D. (2004) Advanced 'tactile' medical imaging for separation surgeries of conjoined twins, in *Childs Nervous System, Monograph on Conjoined Twins*, Spinger-Verlag, Heidelberg, August, **20** (8–9), 547–53. Epub 2004 Jul 22.

Danforth, R. A., Dus, I. and Mah, J. (2003) 3-D volume imaging for dentistry: a new dimension. *J. Calif. Dent. Assoc.*, November, **31** (11), 817–23.

D'Urso, P. S., Hall, B. I., Atkinson, R. L., Weidmann, M. J. and Redmond, M. J. (1999) Biomodel-guided stereotaxy. *Neurosurgery*, May, **44** (5), 1084–93.

Gateno, J., Teichgraeber, J. F. and Xia, J. J. (2003a) Three-dimensional surgical planning for maxillary and midface distraction osteogenesis. *J. Craniofac. Surg.*, November, **14** (6), 833–9.

Gateno, J., Xia, J., Teichgraeber, J. F. Rosen, A., Hultgren, B. and Vadnais, T. (2003b) The precision of computer-generated surgical splints. *J. Oral and Maxillofac. Surg.*, July, **61** (7), 814–7.

Honl, M., Dierk, O., Gauck, C., Carrero, V., Lampe, F., Dries, S., Quante, M., Schwieger, K., Hille, E. and Morlock, M. M. (2003) Comparison of robotic-assisted and manual implantation of a primary total hip replacement. A prospective study. *J. Bone Joint Surg. Am.*, August, **85A** (8), 1470–8.

Miralbell, R., Ozsoy, O., Pugliesi, A., Carballo, N., Arnalte, R., Escude, L., Jargy, C., Nouet, P. and Rouzaud, M. (2003) Dosimetric implications of changes in patient repositioning and organ motion in conformal radiotherapy for prostate cancer. *Radiother. Oncol.*, February, **66** (2), 197–202.

Smith, K. R., Frank, K. J. and Bucholz, R. D. (1994) The NeuroStation – a highly accurate, minimally invasive solution to frameless stereotactic neurosurgery. *Comput. Med. Imaging Graph.*, July–August, **18** (4), 247–56.

Swain, E. (2004) Innovation and good design: a winning combination *Medical Device and Diagnostic Industry*, **26** (4), 53.

Sweeney, R. A., Bale, R., Auberger, T., Vogele, M., Foerster, S., Nevinny-Stickel, M. and Lukas, P. (2001) A simple and non-invasive vacuum mouthpiece-based head fixation system for high precision radiotherapy. *Strahlenther. Onkol.*, January, **177** (1), 43–7.

Worthley, B. W. and Cooper, R. E. (1967) Computer-based external beam radiotherapy planning. I. Empirical formulae for calculation of depth-doses. *Phys. Med. Biol.*, April, **12** (2), 229–40.

Xia, J., Ip, H. H., Samman, N., Wong, H. T., Gateno, J., Wang, D., Yeung, R. W., Kot, C. S. and Tideman, H. (2001) Three-dimensional virtual-reality surgical planning and soft-tissue prediction for orthognathic surgery. *IEEE Trans. Inf. Technol. Biomed.*, June, **5** (2), 97–107.

Yamada, T., Mori, Y., Minami, K., Mishima, K. and Sugahara, T. (2002) Three-dimensional facial morphology, following primary cleft lip repair using the triangular flap with or without rotation advancement. *J. Craniomaxillofac. Surg.*, December, **30** (6), 337–42.

Zelinski, P. (2003) You can automate more than you think. *Modern Machine Shop*, October.

3

Biomodelling

P. D'Urso

3.1 Introduction

Surgery is a practical art! The surgeon is often required physically to intervene to effect a treatment. To minimize operative morbidity and mortality, and to maximize therapeutic success, surgical strategies are tailored to each patient and must be carefully planned using the best possible anatomical information. The traditional way for a surgeon to gain basic experience, without risk to the patient, is to dissect cadavers and to examine carefully preserved pathological specimens. This serves to provide an anatomical and pathological understanding from which operative interventions may be safely made. As every patient is unique, there is a need for the surgeon to attain a specific understanding of the individual's anatomy preoperatively. Thorough physical examination may be all that is needed for common conditions with which the surgeon is experienced. Detailed information displaying the morphology of internal structures is often required by the surgeon to understand more complicated pathological conditions. To obtain this internal anatomical information non-invasively, the surgeon relies on medical imaging.

The discovery of the diagnostic value of X-rays by Wilhelm Conrad Roentgen in 1895 first introduced a way of studying internal anatomy without direct physical intervention. Plain X-rays were quickly accepted for the display of skeletal pathology. The introduction of radio opaque contrast agents enhanced the power of plain X-rays to display the anatomical morphology of soft tissues in many conditions. Major advance came with the introduction of computed tomography (CT) by Sir Godfrey Newbold Hounsfield in 1973 (Hounsfield, 1973). This technology allowed visualization as never before. Neurosurgeons could study direct cross-sectional images of intracranial soft tissue tumour masses. This illustration of neuroanatomy was a great advance compared with angiography, myelography and pneumo-encephalography. As CT became widely used in neurosurgery, applications soon followed in many other specialities. Ultrasound and more recently magnetic resonance (MR) imaging have followed the use of X-ray, and each has come to have specific indications for imaging internal anatomy.

Advanced Manufacturing Technology for Medical Applications   Edited by I. Gibson
© 2006 John Wiley & Sons, Ltd.

These advances in medical imaging have created ever increasing volumes of complex data. The interpretation of such information has become a speciality in itself and the surgeon at times may be left bewildered as to how best to apply the available information to the practicalities of physical intervention. The surgeon seeks to understand the exact morphology of the abnormality, its relationships to surrounding anatomy and the best way to access and correct the pathology operatively. Such specific information is not readily available in the radiologist's report and, however experienced the surgeon may be at interpreting radiological films, it can be difficult to interpret the data so that such questions can be easily answered.

Three-dimensional (3D) imaging has been developed to narrow the communication gap between radiologist and surgeon. By using 3D imaging, a vast number of complex slice images can be combined into a single 3D image which can be quickly appreciated. The term 'three-dimensional', however, is not a truly accurate description of these images as they are still usually displayed on a radiological film or flat screen in only two dimensions.

The advent of 3D imaging has not only dramatically improved data display but also promoted the development of even more useful technologies to assist the surgeon in diagnosis and planning. Ideally in surgery an exact copy of the patient would allow complete preoperative simulation. The yearning of the surgeon for this most realistic portrayal of data initiated the evolution of 3D imaging and has now fuelled the development of solid biomodelling.

'Biomodelling' is the generic term that describes the ability to replicate the morphology of a biological structure in a solid substance. Specifically, biomodelling has been defined by the author as 'the process of using radiant energy to capture morphological data on a biological structure and the processing of such data by a computer to generate the code required to manufacture the structure by rapid prototyping apparatus' (D'Urso and Thompson, 1998). A 'biomodel' is the product of this process. 'Real virtuality' is the term coined to describe the creation of solid reality from the virtual imagery (D'Urso and Thompson, 1998). Virtual reality, in contrast, creates a computer-synthesized experience for the observer without a real basis. In medicine, biomodelling has been used to create anatomical real virtuality. Biomodels are a truly remarkable and exciting tool in the practice of surgery, and the applications of such a generic technology will be discussed in this chapter.

3.2 Surgical Applications of Real Virtuality

The author initiated research into biomodelling technology in 1991. Anatomics™ Pty Ltd (www.anatomics.net) was subsequently founded as a research and development company in 1996. This chapter will reflect the author's experience with over 600 patients where biomodelling technology has been applied in a broad range of applications. Such applications have been developed in cranio-maxillofacial surgery, neurosurgery, customized prosthetics, orthopaedics and other miscellaneous specialties. The breakdown of applications is listed in Table 3.1.

In general, biomodels were found to have five general applications. These were first identified in craniofacial surgery (D'Urso *et al.*, 1998) but also apply as generic medical applications of the technology:

- for surgical team communication, to educate patients and improve informed consent;
- to assist surgeons with diagnosis and surgical planning;
- for the rehearsal and simulation of surgery;
- for the creation of customized prosthetics;
- for the accurate placement of implants.

Table 3.1 Basal ganglia

Speciality	Cases
Cranio-maxillofacial	
Craniofacial	99
Maxillofacial	216
Custom prosthetics	163
Orthopaedic	
Spinal	42
Pelvic	10
Upper extremity	2
Lower extremity	6
Neurosurgical	
Skull-based tumour	11
Stereotactic	8
Vascular	31
Basal anglia	2
Miscellaneous	
Aortic	3
Ears	13
Bile ducts	2
Chest	2
Fetus	3
Natural sciences	20
Total	**633**

3.2.1 Cranio-maxillofacial biomodelling

The complexity of cranio-maxillofacial anatomy combined with the morphological variation encountered by the reconstructive surgeon makes cranio-maxillofacial surgery a conceptually difficult task in explanation, planning and execution. The need for clear morphological understanding in such surgery played a large part in the development of both 3D imaging and solid biomodelling. The ideal method of displaying cranial anatomy is the patient's dried skull itself. The biomodelling system developed by Anatomics™ has now come very close to achieving this.

Cranio-maxillofacial surgeons have shown a high level of affinity for biomodelling, perhaps because they routinely use dental casting and the articulated models produced from such castings. The use of biomodels of the mandible and maxilla is a natural extension to the use of these dental castings. The biomodels have been commonly used to improve the diagnostic relevance of the data and for surgical simulation. A highlight of the use of biomodels in cranio-maxillofacial surgery has been the ready acceptance of biomodelling by the surgeons (D'Urso *et al.*, 1998; Arvier *et al.*, 1994; D'Urso *et al.*, 1999a). Consequently they have developed many interesting applications.

In craniofacial surgery, biomodels have been traditionally used by surgeons to gain insight into unusual or particularly complex cases. Often in craniofacial surgery a multidisciplinary team is involved. The biomodels are often used to assist communication among team members, to achieve informed consent from patients and relatives and between surgeons intraoperatively. Biomodels have also been used to simulate craniofacial reconstruction. Standard surgical drills and saws can be used to fashion osteotomies. The bone fragments can be reconstructed using wire, plates and screws or glue. A technique that has been developed to assist with the reconstruction involves the manufacture of two biomodels (D'Urso *et al.*, 1998). The first biomodel is reconstructed by surgeons and acts as an 'end-point' biomodel, illustrating the desired preplanned reconstruction. Surgeons then mark on the second biomodel the osteotomy lines that they intend to use. This biomodel becomes the 'start-point' biomodel. Intraoperatively the two biomodels are used in different ways. The start-point biomodel is used to navigate the anatomy and accurately to transfer the osteotomy lines onto the exposed patient's skull. The osteotomies are then made and the pieces of bone reconstructed according to the end-point biomodel. Such reconstruction, as well as the shaping of bone grafts, can be performed by a second surgeon working at a side table while the other surgeon continues to operate on the patient. This technique has been reported to improve the results of surgery as well as shorten the operating time. The end- and start-point biomodels have also proven very useful for informed consent and team communication.

Maxillofacial surgeons have also developed some interesting specific applications of biomodelling that warrant further discussion (Arvier *et al.*, 1994). These are as follows:

- integration of biomodels with dental castings to form articulated jaws;
- the use of biomodels to shape prosthetic and autograph implants;
- the use of biomodels to prefabricate templates and splints;
- the use of biomodels in restorative prosthetics;
- the use of biomodels to plan distraction osteogenesis.

3.2.1.1   Integration of biomodels with dental castings

The combination of dental casting (replicating the teeth) with biomodels (replicating the jaws) may be advantageous for the following reasons:

- The presence of metal artefact from dental fillings and braces is translated into the biomodel which can result in unacceptable inaccuracies. CT scanning in planes horizontal to the artefact source can minimize the number of planes in which image distortion occurs but may still not be sufficient.
- The accuracy of CT scanning even at high resolution may not be sufficient for the biomodelling of opposing teeth to form a truly accurate occlusive bite.

Traditional plaster dental casting creates a highly accurate, relatively cheap model of the teeth. A combination of these castings with a biomodel mandible and maxilla forms a highly accurate hybrid with considerable benefits. Such a hybrid biomodel may be made by creating a wax key to localize the mandibular dentition in relation to the mandibular body (Arvier *et al.*, 1994). The dentition is then cut away and the key used to locate the plaster casting in relation to the mandible biomodel. In a similar way the upper dental cast may be combined to the maxilla. A temporomandibular articulator may then be created by accurately occluding the dental arches about the temporomandibular joint. Such an articulated biomodel is particularly helpful in orthognathic surgery where the effects of osteotomies on the dental occlusion may be preoperatively examined.

Another way to avoid the difficulties associated with biomodelling the teeth and jaws is to use a laser scanner to create a surface file from the dental casting and then superimpose the file with data derived from CT. In this way an accurate dataset describing the teeth and jaws may be generated that could be used directly to biomodel the teeth and jaws from SL. Such a biomodel would accurately reproduce the dentition as the laser scanning may be performed with far higher resolution than CT scanning. The difficulties with this approach are the accurate co-registering of the two differing datasets in the correct anatomical position and acquisition of dental data that may be inaccessible to the laser scanner.

3.2.1.2   Use of biomodels to shape maxillofacial implants

Biomodels may be used in several ways to shape maxillofacial prosthetic and autographic implants. A simple way to do this is to use the biomodel as a template on which a graft may be directly shaped intraoperatively (D'Urso *et al.*, 1998; Arvier *et al.*, 1994; D'Urso *et al.*, 1999a). The bone graft may be harvested from the iliac crest and shaped directly on the sterilized biomodel. Once the contouring is satisfactory, the surgeon places the graft *in situ* and fixes it. This approach can dramatically reduce operating time while improving the end result. This can be achieved because the surgeon can shape the graft on the biomodel while the assistant simultaneously prepares the exposure of the donor site. This technique also avoids the need for repeated 'fitting and chipping' of the graft when the patient is directly used as the template, since direct shaping is restricted by soft tissue cover and limited surgical access.

Another approach is to use acrylic, or a similar material, preoperatively to create a master implant to serve as a guide for the shaping of the bone graft intraoperatively. This is particularly appropriate when the graft requires a somewhat complex shape. The surgeon can minimize operating time by preoperatively moulding the acrylic to the exact shape required, using the biomodel as the template.

More recently, synthetic bioabsorbable materials, such as polygalactic acid, have been introduced into cranio-maxillofacial surgery. Such materials in the form of sheets or plates can be intraoperatively shaped to fit specific anatomy. As with autograph implants, such bioabsorbable implants can be shaped to a biomodel to save time and avoid difficulties with limited anatomical exposure.

3.2.1.3   Use of biomodels to prefabricate templates and splints

Biomodels may be used to plan endosseous surgery and to create customized drill guide templates (Arvier *et al.*, 1994). Edentulous patients may have teeth restored by mounting them on titanium pins which are implanted into the jaw. The implantation of the titanium pins, however, can be difficult and can be complicated by damage to the underlying dental nerve. Mandibular biomodels accurately replicate the neurovascular canal through which the mandibular nerve travels. The course of this nerve may easily be displayed by passing a malleable coloured wire along the neurovascular canal or replicated in a second colour using StereoCol resin. The biomodel can then be used to determine and rehearse the positioning and depth of the holes required to receive the titanium mounting pins using a standard dental drill. The pins can then be inserted into position and 'cold-cure' acrylic moulded around them and the mandibular contour to form a relocatable drill guide. The depth of each hole can also be determined relative to the drill guide and recorded. During the surgery the mucosa is stripped from the mandible and the drill guide matched using the reciprocal contours. While firmly held in place, the guide can be used to drill the holes with the correct positioning and depth as preplanned in the biomodel without risk of injury to the underlying mandibular nerve. More recently, SimPlant software has been used preoperatively to simulate the placement of implants in a virtual enviroment. Rapid prototyping is then used to generate a biomodel of the jaw as well as a custom drill guide template (Vrielinck *et al.*, 2003).

Dental splints may be prefabricated using articulated biomodels of the teeth and jaws. Such splints may be useful to maintain the relative position of the dental arches after osteotomy surgery. The surgery is rehearsed on the biomodel and the relative position of the bones determined and set. A splint may then be moulded to fit to the reconstructed biomodel. At the end of surgery the splint may then be used to maintain positioning. This technique can save the time taken at the end of the procedure to mould such a splint. The risk of bony movements while moulding the splint directly is also avoided with this technique (Arvier *et al.*, 1994).

Distraction osteogenesis is a technique used to promote the remodelling and lengthening of bones. The technique uses an implantable device slowly to distract an osteotomized bone by about 1 mm per day. Biomodels have been used to plan the positioning of the distraction device as well as to determine the extent of distraction to achieve the desired result. The use of biomodels has proven extremely valuable in this regard (Yamaji *et al.*, 2004; Whitman and Connaughton, 1999).

3.2.1.4 Use of biomodels in restorative prosthetics

Yet another use of biomodels in maxillofacial surgery is for the prefabrication of restorative prosthetics. Surgery usually requires titanium fixative implants to be inserted on which a prosthesis is mounted. A biomodel can be used to determine the best location for these implants and to construct the overlying prosthesis. This may be performed by inserting the mounts into the biomodel, constructing the overlying wire scaffolding and then adding the plastic nasal and dental prostheses. A biomodel is therefore used not only to plan and rehearse implantation but also for the construction of the prosthesis. This improves the ability to form an accurate prosthesis, which enhances the cosmetic result and shortens the operative time. Another advantage is that less time is required to shape and trial fit the prosthesis, as much of this can be performed on the biomodel.

Biomodelling has also been used to generate a prosthetic ear replacement. CT scanning is performed and the normal ear is mirror imaged. A biomodel is then used as a master to cast a synthetic substitute ear which is fastened via osteointegrated titanium screws.

3.2.2 Use of real virtuality in customized cranio-maxillofacial prosthetics

The sight of unfortunate patients with disfiguring skull defects on the neurosurgical and rehabilitation wards has always been disturbing. These defects are not only cosmetically displeasing but also pose a significant risk to the patient's underlying brain should trauma occur. The method traditionally used by neurosurgeons, before the advent of biomodelling, to repair craniotomy defects (for which no autologous bone is available) is to shape cold-cure acrylic to fit the defect. This acrylic is moulded and polymerized *in situ* to form the implant. The technique is limited by the artistic skill of the surgeon to achieve the desired contour of the implant within the short time before polymerization. This moulding process can be time consuming, especially if the surgeon takes several attempts to achieve the desired contour. Longer surgical time increases infection risk. Infection is a major complication in any implant surgery as its presence will usually necessitate removal of the implant. Another disadvantage of cold-cure acrylic monomer is its toxicity. If polymerization, is incomplete, monomer can leach into the patients tissues with detrimental effects.

The application of real virtuality to assist in the manufacture of customized prostheses was realized at an early stage (D'Urso *et al.*, 1994). Early techniques were based on the creation of CNC milled models that could be made into a mould from which the implant could be cast (White, 1982; Toth, ellis and stewart, 1988). A model could also act as a template. Wax could be moulded to create a master implant that could then be used to create a mould from which the implant could be cast. Alternatively, the model could act as a template over which titanium (Blake, MacFarlane and Hinton, 1990; Joffe *et al.*, 1992; Abbott *et al.*, 1994) or hydroxylapatite (Waite *et al.*, 1989) could be moulded to generate an implant. Such techniques have also been applied to construct prostheses in maxillofacial surgery and orthopaedic surgery (Rhodes *et al.*, 1987).

New methods, using biomodelling, to construct cranioplastic prostheses have been developed. A master is generated by a computer graphic mirroring process whereby the normal side of the cranium can be manipulated to produce an exact contour of the 'missing part' of the skull. This 'missing part' and the biomodel of the craniotomy defect thus form both the male master implant and the female biomodel of the patient's skull. The SL master implant can be hand finished exactly to fit the biomodel of the craniotomy defect, which has a high degree of accuracy (D'Urso *et al.*, 2000; Barker, Earwaker and Lisle, 1994). Any small irregular features around the edge of the master implant are smoothed to create a fit with the biomodel achieving maximum contact. The SL master implant may then be used to create a mould from which an acrylic implant is cast.

If the defect extends beyond the mid-line, several techniques can be used to generate a master implant. Computer interpolation from existing anatomy can be used to fill the deficit. The superimposition of a standard skull dataset with the patient's skull can be performed. The patient's anatomy is then subtracted to leave a master implant. The simplest and most effective technique involves the direct sculpting of a wax master to fill the defect.

The use of rapid prototyping to manufacture both male and female prosthetic components simultaneously has significant advantages over previous techniques:

- it negates the need for wax sculpting of the master;
- it directly provides a master implant;
- it can be used with all known prosthetic materials.

Methyl methacrylate (acrylic) has been long accepted for use in cranioplasties (Manson, Crawley and Hoopes, 1986; Remsen, Dawson and Biller, 1986). Other materials allow the ingrowth of bone, such as hydroxylapatite (Waite *et al.*, 1989), ceramic (Kobayashi *et al.*, 1987) and ionomeric bone cement (Ramsden, Herdman and dye, 1992). These materials are highly expensive, somewhat difficult to mould or shape and, if infected, can be difficult to remove *in toto*. Titanium is also used commonly (Eufinger *et al.*, 1995; Blake, MacFarlane and Hinton, 1990; Joffe *et al.*, 1992; Abbott *et al.*, 1994) and has the advantage of being biologically inert, although this is offset by cost, difficulties in moulding, casting or milling and the artefacts generated in CT and MR after implantation.

3.2.2.1 Computer mirroring techniques for the generation of prostheses

Mirror imaging techniques have been applied frequently to assist the assessment of asymmetrical deformities of bone and soft tissues in the cranio-maxillofacial region. In order to isolate the dimensions of a missing part of anatomy, digital subtraction mirror imaging may be used (Fukuta and Jackson, 1992). In this technique the CT data are segmented to isolate the tissue of interest, usually bone. The original image is saved. A mirror image of the image about the sagittal plane is then created and its abnormal half erased. The normal half of the original image is also erased. The original image (containing the missing anatomy) is then subtracted from the normal mirrored image. This should leave the missing anatomy alone, which can then be used to generate a 'master' implant. The master implant data and original data are then used to manufacture biomodels. Alternatively, the master implant data may be used directly to fabricate the implant. CNC milling of titanium or the use of selective laser sintering have been reported in this regard.

As the mirroring technique is essentially a two-dimensional approach to a three-dimensional problem, the symmetry of data acquisition from the patient is critical. CT scanning must be performed without gantry tilt and patient movement, and with isoaxial positioning of the patient. These criteria are difficult to attain in routine scanning as it is extremely difficult to place the patient's head in the scanner without slight rotation or movement occurring. Even one or two degrees of misalignment is significant. The plane of symmetry in the images is especially difficult to determine when the routine reference landmarks such as orbit or nose are deformed. Natural asymmetry may complicate digital subtraction mirror imaging in any particular patient. Each slice needs to be checked by the operator to determine the influence of the natural asymmetry, and allowances and alterations may be necessary to accommodate for this which can become time consuming. If rotation of the patient occurs during CT scanning it is possible to reformat an asymmetrical CT scan. Complex algorithms have been developed to solve many of these issues, and the virtual creation of the 'missing' anatomy is possible but often time consuming and limited by possible error.

It should also be noted that any inaccuracies created in a master implant can be readily corrected by burring away material and/or adding wax to it so that it fits exactly into the craniotomy defect reproduced in the biomodel. In this way, hours of computer work to attain an accurate master can be accomplished within minutes by hand in the workshop. This observation highlights the utility of the solid master implant compared with the virtual master implant. The generation of a master implant has another advantage in that a thoroughly validated material, such as acrylic, can be used to fashion the final prosthesis.

In our experience the best method for the generation of customized implants is for a prosthetist to sculpt the master from wax or directly from titanium on the biomodel itself (D'Urso *et al.*, 2000). This allows for subtle anatomical variations to be incorporated as well as allowing for the surgeon's input in the customization process. Again, the use of real virtuality appears to have more utility than a virtual alternative.

Once generated, the implant is sterilized by gas or autoclaving. The implant is used by the surgeon in conjunction with the biomodel of the region to determine the exact attachment sites and means. Attachment is usually achieved with titanium miniplates and screws. Titanium miniplates and screws can also be preoperatively attatched to the implant to minimize operating time.

3.2.2.2   Results of implantation

In our experience of over 160 implants manufactured by Anatomics™, all have fitted well except for one. The implants required only minor trimming to achieve good contact between the bevelled edge of the implant and the bone defect. The fit after such minor trimming was consistently within 1.5 mm. The use of prefabricated implants was reported to reduce operating time compared with the traditional cold-cure method. Surgeons have reported that the ability to study the implant and the biomodel preoperatively allows fixation to be exactly planned and possible problems to be identified, e.g. the site of cranial sinuses to be determined (D'Urso *et al.*, 2000).

All the patients were happy with the cosmetic result, and several commented that they could not even tell that a cranial implant was present. The patients were particularly pleased to have the opportunity to examine their implant before surgery. The implant and biomodel were helpful in obtaining informed consent.

Only one surgeon reported that significant trimming of around 5 mm was required to fit the implant to a large frontal defect. It is not clear why this happened. The most likely cause was that the biomodel warped after being removed from the SLA machine or that the surgeon failed to remove all of the soft tissues from around the craniotomy defect. Warpage has been reported as a source of error by other investigators.

3.2.2.3   Advantages of prefabricated customized cranioplastic implants

The advantages of the biomodel-generated cranioplasties compared with the traditional techniques used for cranioplastic surgery where autologous bone is not available can be summarized as follows:

- improved cosmesis and fit;
- non-toxic thermally polymerized acrylic used;
- reduced operating time and risk of infection;
- opportunity for surgeons to assess the defect and implant preoperatively, and improve fixation;
- improved patient informed consent;
- allows one-stage resection and implantation procedures to be performed.

3.2.3   Biomodel-guided stereotaxy

Stereotactic (Greek *stereo* 3D, and *tactic* to touch) surgery can be defined as the ability accurately to localize brain structures after locating the discrete structures by means of an apparatus co-registered with 3D coordinates. Biomodel-guided stereotaxy uses a physical phantom as the source of the 3D coordinates for the precise localization of intracranial anatomy. This original technique truly reflects the Greek origins of the word *stereotactic*. As this application of biomodelling is novel, the background of stereotaxy will be discussed before detailing the experiments that were conducted.

3.2.3.1 Development of stereotaxy

Prior to the development of CT, stereotactic techniques were based on head frame systems calibrated to the anterior–posterior commissural (AC–PC) line visualized from intraoperative ventriculography (Anichkov, Polonsky and Usov, 1977). The advent of CT enabled the development of the 'modern art' of stereotaxy. CT digitally displayed anatomy in detail and with only minor geometrical distortion. By combining CT with a localizing frame and fiducials, Brown (Brown, 1979) developed CT stereotaxy. The fiducial system serves as a reference by which any target on a CT/MR scan can be defined in terms of its location with respect to the head frame.

Frame-based stereotaxy is disadvantaged by the traumatic attachment of a head frame, cluttering of the operative field and limited application of evolving 3D imaging techniques. As many neurosurgical procedures do not require pinpoint accuracy, the inconvenience of the frame is often not warranted. The need for a more ergonomic system that easily integrates the anatomical information derived from 3D imaging has spawned the development of frameless stereotaxy. Three or more small stereotactic fiducials are fixed to the patient's scalp prior to volumetric CT or MR. The data are transferred to a operating room workstation. Infrared light projected from a mounted emitter/receiver device is used to localize a wand with reflective fiducials in relation to the patient fiducials. The triangulation of the transmitter positions and the wand tip allows the co-registration of the wand position and the 3D image derived from scan data. This sophisticated apparatus and 3D computer software thus allow the surgeon surgically to navigate with the wand by visualizing its position in relation to the image data.

Sources of error inherent in all frameless stereotaxy systems are related to the quality of image data, the localizing apparatus, registration error (caused by movement of fiducials attached to skin or movement of skin surface landmarks), the deformation of soft tissues during scanning, operator error and movement of the brain relative to the markers or 'brain shift' (caused by craniotomy, tumour resection, CSF aspiration, brain swelling and retraction). Such a system is also expensive and complicated and requires significant training and know-how.

3.2.3.2 Development of biomodel-guided stereotactic surgery

Biomodel-guided stereotactic surgery uses a solid replica of the patient (biomodel) to provide the anatomical data needed for trajectory planning. This concept forms the basis of a method of stereotaxy that can be performed using a number of different apparatus (D'Urso, 1993).

The method is as follows:

1. The patient undergoes CT scanning and a biomodel is manufactured.
2. An apparatus is attached to definable points on the replica.
3. The apparatus is used to create an operative plan so that interventional coordinates are able to be saved (e.g. a trajectory is defined to localize a target).
4. The apparatus is transferred and attached to the patient via common definable points.
5. The intervention is repeated on the patient using the saved coordinates.

This original method has been validated by the use of apparatus in phantom and patient studies.

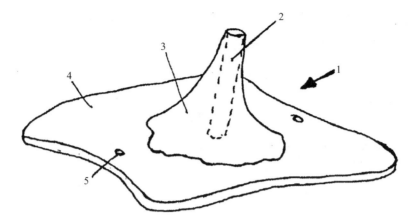

Figure 3.1 *1* Stereotactic template; *2* trajectory barrel; *3* moulded support; *4* contoured base plate; *5* localizing holes to marker points

3.2.3.3 Biomodel-guided stereotactic surgery with a template and markers

The surgeon draws the resection margin of the tumour on a biomodel of the patient (see Figure 3.4). The resection is performed on the biomodel. A custom-made template is made to fit the operative plan on the biomodel by moulding acrylic to the biomodel so that its outer boundary replicates the surgeon's original marked resection. The resection is rehearsed and the biomodel left with a large defect. The template is used to guide the resection on a mirrored biomodel of the cranium. The 'normal' resected plate is then used as a master implant and is hand finished and used to mould and cast an acrylic custom implant.

Intraoperatively, the surgeon places the template over the skull and contour localizes it. The template is held firmly to the skull and the resection margin is traced from the boundary of the template onto the patient's cranium using diathermy. The tumour is then resected using the traced margin as previously rehearsed on the biomodel. The preoperatively customized acrylic implant is used to close the defect.

The transfer template to perform biomodel-guided stereotaxic surgery may take several forms, each with its own merits:

1. A simple *impression template* may be produced. Moulded plastic can be used to trace out a boundary or to guide an intracranial trajectory. Such a template can be fabricated on the biomodel and then be transferred to the patient and aligned by the matching of surface contours alone. This simple device may be difficult to localize if the contour is not sufficiently distinctive. This technique was mentioned previously for the placement of endosseous implants.
2. *Surface markers* can be used to locate a template precisely. The template is located to surface contour and at least two markers. The template may be used to trace a boundary for a craniotomy and preoperatively customize an implant to close the defect. One or multiple intracranial trajectories may be incorporated into the template by means of barrels. The template may be made to encase multiple implants by means of a lid with screws. This technique has special application for interstitial brachytherapy (Poulsen *et al.*, 1999).

The accuracy of these experiments was comparable with the previously described stereotactic techniques.

3.2.3.4   Biomodel-guided stereotactic surgery using the D'Urso frame

Frame description

The desire to improve the apparatus used for biomodel-guided stereotactic surgery led to the development of a new stereotactic frame (Figure 3.2). A prototype was made using SL and, after design modifications, a stainless steel frame was manufactured. This D'Urso stereotactic frame has been designed to be usable, simple and accurate. The device is ergonomic in the surgical practice of biomodel-guided stereotactic surgery. The reusable frame has three legs which allow exact marker point attachment and a ball-in-socket instrument guide which affords a wide trajectory angle. Adjustable feet easily attach to surface marker pins.

To validate the accuracy of the D'Urso frame, a phantom study was performed. The mean measurement error between the biomodel trajectory and the skull target was 0.89 mm with a 95% confidence interval between 0.86 mm and 0.93 mm (D'Urso *et al.*, 1999b).

Technique

The patient is examined by the surgeon and the best position of the frame is determined using standard imaging data or previous biomodels see Figure 3.5. The frame is then held against the patient's head and the scalp where the feet of the frame rest is shaved and infiltrated with local

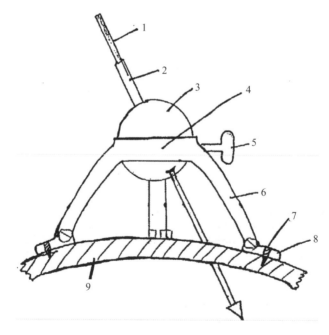

Figure   3.2 Diagram of the original D'Urso frame: *1* instrument in barrel; *2* trajectory barrel; *3* pivot ball; *4* base plate; *5* lock nut to hold trajectory; *6* tripod base; *7* marker pin; *8* adjustable foot; *9* cranial surface

anaesthetic. The frame is held firmly against the patient's head and is used to guide the insertion of three marker pins into the bone cortex or onto the scalp. Non-ferrous multimodality fiducial imaging markers made to minimize the generation of artefact on CT and MR are available. The frame is then removed, the patient undergoes CT scanning and a biomodel is created in the standard way.

The patient is prepared for surgery in the usual fashion. The sterilized frame is attached to the marker pins replicated on the sterilized biomodel. The trajectory and depth for the procedure is determined and the frame is locked. The locked frame is then transferred to the patient's head and attached to the exposed marker pins. The frame then acts firstly as a drill guide to breach the patient's cranium and then as a trajectory guide for the procedure. The locking nature of the frame allows the precise replication of the coordinates selected on the biomodel. Alternatively, the settings of the frame can be recorded from calibrations and then reset prior to localization on the patient.

The frame can be used easily for multiple trajectory transfer from model to patient, for multiple biopsies or for marking a preplanned resection margin.

3.2.3.5 Utility of biomodel-guided stereotactic surgery

This new technique of biomodel-assisted stereotaxic surgery has advantages over presently used methods of stereotaxic surgery for the following reasons:

1. Simple mechanical methodology, based on rehearsal of the procedure on an accurate phantom, reduces the degree of system and surgeon error.
2. Little training is necessary, as the surgical skills required are mechanical and the same as those used in standard operations.
3. The system is user friendly, requiring no computer or computations to implement in the operating theatre.
4. The D'Urso frame is simply designed, cheap and reusable and the only system cost. Biomodels are made on a case-by-case basis.
5. The patient can be scanned without the frame attached several days before the procedure.
6. Realistic and accurate biomodels provide a readily recognizable solid replica of a person's anatomy that may be used for tactile preoperative planning.
7. Biomodels may be used realistically and interactively to simulate surgery, i.e. the stereotaxic trajectory. This allows the surgeon to plan modifications to standard techniques and view the planned result.
8. Biomodels provide patients with a clearer understanding of their pathology and the aims and limitations of surgery preoperatively, improving informed consent.
9. Craniotomies and resections may be preplanned, rehearsed and transferred to the patient. Closure is with a customized preoperatively fashioned implant.
10. Biomodels require no specialized equipment or knowledge for interpretation and use, are rugged and may easily be transported and sterilized for intraoperative use.
11. The D'Urso frame is small and light and can be used intraoperatively as desired, allowing clear surgical access.
12. The D'Urso frame is accurate regardless of the patient's head position, and is not affected by intraoperative movement of the head.

13. The accuracy of the D'Urso frame is not affected by external interference from light, acoustic or electromagnetic sources.
14. Once the patient is marked, the D'Urso frame can be used without further calibration on multiple occasions at any time.
15. The D'Urso frame is not affected by error generated from distortions of the patient's scalp during scanning, or before and during the operation.

These studies have been carried out at low cost with relatively simple materials. Advances in biomodelling technology will produce biomodels with near-perfect accuracy, coloured soft tissue, blood vessels and bone at low cost. These advances will allow superior results from biomodel-guided stereotaxic surgery.

The main drawbacks of biomodel-guided stereotaxy as described are:

1. The need for the manufacture of a SL biomodel which takes <24 h and presently costs <A$1000. The manufacturing time is not so important as most stereotactic procedures are non-urgent and can be scheduled accordingly. The cost will come down with time as biomodelling technology becomes more widespread.
2. The need for invasive markers to be attached to the patient's cranium. Non-invasive markers can be used but the accuracy may be reduced.
3. The difference in size between the biomodel and patient of less than 1 mm is another problem with this system. This error is mainly due to that generated in the scanning process and could possibly be reduced by introducing a scaling factor into the imaging software. Improvements in scanning and software should be able to minimize biomodel inaccuracy to <0.25 mm, which would significantly improve the accuracy of biomodel-guided stereotactic surgery.

3.2.4   Vascular biomodelling

The investigation of the feasibility of vascular biomodelling has been an important element of biomodelling research for the following reasons:

- The biomodelling of delicate cerebral vessels would clearly demonstrate the ability of the process to replicate fine complex structures. This capability is not immediately apparent when biomodelling larger structures such as the cranium. Biomodels of blood vessels would establish the quantum leap that the technology has made in comparison with CNC milling techniques.
- The ability to incorporate blood vessels in biomodels greatly enhances their value in stereotaxy and skull-base tumour surgery. It increases the diagnostic utility and allows the surgeon better to plan safe surgical access.
- Vascular biomodels will open applications in vascular surgery (e.g. cerebral aneurysms) and for the development of customized prosthetic grafts (e.g. aortic).

3.2.4.1   Biomodelling from CTA

The feasibility of vascular biomodelling was first assessed by experimental work with the CT angiogram of a patient's circle of Willis which contained an anterior communicating artery aneurysm (D'Urso *et al.*, 1999c). The thresholds for bone and vessels are similar but not identical. Cortical bone tends to have a higher threshold and, depending on contrast concentration, the vessels a somewhat lower threshold. If the threshold is set too high then the finer vessels are lost, and if the threshold is set too low then dense soft tissues close to the bone are included as bone. Various techniques have been utilized with BioBuild™ accurately to define both the bone and the vasculature. A connectivity algorithm was used on only the larger vessels in the circle of Willis so that unconnected smaller branches could be deleted. Smaller branches tend to become disconnected as the density of contrast within falls below the original threshold and as their diameter becomes such that partial volume effects create gaps in the data. The result of thresholding and connectivity was used to generate data for biomodelling.

The most difficult part of the study was removing the supports. Great care must be taken when removing the supports to avoid damage to finer vessels. The finished biomodel was very good, with clear resolution of the entire circle of Willis and main vessels down to 1 mm in diameter. The aneurysm arising from the anterior communicating artery was well replicated with a 3 mm diameter. Both arterial and venous vessels were included which at first appearance could be confusing but, if aided by good anatomical knowledge, provided a clear accurate representation. The quality of the biomodel was such that even smaller aneurysms could be replicated.

3.2.4.2 Biomodelling from MRA

MRA data from a 1.5 Tesla General Electric scanner has been used to biomodel cerebral angiograms containing aneuysms (D'Urso *et al.*, 1999c). The main difference between MRA and CTA data is that the MRA arterial structures are of much higher pixel intensity in the image and are thus easily segmented. The biomodels replicated fine vessels with a diameter of 0.8 mm. The resolution of MRA was noted to be less than that of CTA.

3.2.4.3 Clinical applications of vascular biomodels

Cerebrovascular biomodels have been found to be very useful to surgeons for preoperative evaluation and intraoperative reference (D'Urso *et al.*, 1999c; Wurm *et al.*, 2004). Neurosurgical applications that have been identified include surgery for intracranial aneurysms and arterio-venous malformations. For these conditions the biomodels were found to represent accurately the vasculature at operation. Neurosurgeons reported that the biomodels gave a better overview of the anatomy compared with standard 2D and 3D images as the anatomy is often complex and difficult to interpret from multiple 2D projections which are often selected by the radiographer. The biomodels allowed the surgeons physically to manipulate the data into any perspective. Biomodels were sterilized and used intraoperatively for microsurgical navigation. The biomodels allowed the surgeon to anticipate problems and assisted in the selection of appropriate aneurysm clips. The biomodels were also very valuable for gaining consent from patients and relatives. As cerebrovascular surgery carries a considerable risk of morbidity and mortality, the biomodels were seen to be particularly cost effective. It was noted that the small resin volume in most cerebrovascular biomodels meant that they could be produced at low cost.

Several biomodels of abdominal aortic aneurysms were also manufactured. These biomodels were produced to ascertain feasibility rather than for specific surgical use. Other authors have also reported the manufacture of aortic biomodels (Lermusiaux *et al.*, 2001). The introduction of multislice helical CT will allow longer vessels to be manufactured, and it is likely that such biomodels will be useful for planning the endovascular repair of diseased aortic, iliac and renal arteries (Kato *et al.*, 2001).

Another interesting vascular application of biomodelling is the biomodelling of the heart and its valves. The replication of abnormal heart valves may assist in the study of the effects on haemodynamics (Gilon *et al.*, 2002). Work has also been done to assess cardiac development (Pentecost *et al.*, 2001) and the use of echocardiography as a data source (Binder *et al.*, 2000).

3.2.4.4 Vascular biomodelling: technical note

A factor to consider when viewing vascular biomodels was that the vasculature contained was only a representation of the internal lumen. The relatively constant oversize ($\sim$0.7 mm) of the biomodels demonstrated by the earlier accuracy study would, however, compensate for the vessel wall and provide an accurate replica.

We identified some difficulties in vascular biomodelling. The most significant problem is the editing of unwanted vessels. CTA and MRA have the ability to resolve fine vessels to a diameter of <0.5 mm (as does stereolithography) which if contained in the biomodel would require a confusing network of supports. Post-processing would be difficult and lead to either the inclusion of supports within the finished biomodel or the destruction of fine vessels as the supports were removed. The steps that were used to eliminate these vessels worked well to preserve the larger structures but may lead to the deletion of important vasculature if performed without a good understanding of the region of interest within the dataset.

Another difficulty when biomodelling vessels and bone together (from CTA) is the difficulty in defining the vessel as it passes through bone. In CTA the subtle difference in the density of the bone and vessel makes thresholding very difficult. Two-colour StereoCol resin is best used to highlight the internal course of a vessel through the skull base.

Inherent individual benefits and drawbacks of CTA and MRA were noted. CTA by combining bone allows for a careful evaluation of an aneurysm in relation to bony anatomy, thus assisting the surgeon with the planning of the surgical approach. CTA, however, has the issue of thresholding bone and vessel and difficulties in differentiating venous from arterial structures.

MRA has the advantages of displaying only arterial structures, free of bony artefacts, without the need for contrast but the disadvantage of not having bony support or landmarks. This can make the MRA biomodels delicate and difficult to orient. MRA data also have lower resolution and any biomodel generated is likely to reflect this.

Future work is being directed toward the use of 3D rotational digital subtraction angiography (Wurm *et al.*, 2004). This imaging technique uses multiple 2D projections of subtraction angiography to reconstruct a high-resolution 3D vascular model. The use of these data for biomodelling is likely significantly to improve the resolution of the blood vessels as well as avoid artefacts associated with venous and bony structures.

3.2.5 Skull-base tumour surgery

Tumours occurring in the region of the cranial floor occupy the most anatomically complex and inaccessible part of the human body. The anatomy of the cranial floor leaves little room for error and allows only limited surgical access. Resection or debulking of these tumours is immensely challenging for the surgeon and of high risk to the patient. To minimize operative morbidity and mortality, and to maximize therapeutic success, accurate surgical strategies must be tailored to each patient and must be carefully planned using the best possible anatomical information. To optimize data display, several authors have applied 3D CT technology to skull-base tumours (Vannier, Gado and Marsh, 1983; Gillespie, Adams and Isherwood, 1987; Howard, Uster and May, 1990). This technique was found to simplify interpretation, enhance communication between the radiologist and clinician and facilitate surgical planning. The use of solid skull-base tumour biomodels had not previously been reported but seemed to be a logical application of biomodelling. Biomodelling skull-base tumours also provided the opportunity to combine bone, soft tissue and vasculature (D'Urso et al., 1999d).

The biomodelling of skull-base tumours is challenging because two or three (with blood vessels) differing tissue types need to be incorporated into the biomodel. The differing densities of the soft tissue composing the tumour and the adjacent bone must be edited from all other structures. When only biomodelling the bone and vasculature, this is achieved by the relatively simple process of segmentation. The voxel threshold can be set so that only bone and vessels remain (high threshold), removing all of the soft tissues (lower thresholds). The problem encountered when biomodelling a soft tissue mass is how to separate it accurately from surrounding structures of similar density. Fortunately, tumours tend to be more vascular than the surrounding normal tissues, and contrast agents will highlight them in the image data. This enhancement in the threshold values of tumour voxels can be used to edit the tumour by segmentation, but not to the extent that it can be relied on completely to separate the abnormal from normal. The enhanced threshold due to contrast is still somewhat below that of cortical bone so that, if the bone threshold is used, the tumour will still be lost. In order accurately to include the tumour and bone together, it is necessary to edit the data on a slice-by-slice basis. The threshold is set just below that of the contrast enhanced tumour, and on each image slice a pixel editor is used to trace the boundary of the tumour and internally fill it with a threshold equal or above that of bone. In this way, a single threshold can be used to separate the tumour and bone from all surrounding soft tissues (D'Urso et al., 1999d).

This technique requires a steady hand and good mouse control to accurately move the on-screen cursor when tracing the tumour boundary. The use of a pen on an editing image pad has been found to be a much better interface to ensure the required accuracy.

Another problem with this technique is that the single high threshold used to paint the tumour boundary required for segmentation creates a steep threshold gradient between the tumour and surrounding soft tissue which results in a somewhat stepping boundary after interpolation of the data. This is because the interpolation algorithms average a range of thresholds representing a gradual change from one tissue to another to estimate where the segmentation should occur, giving a smoothed appearance to the biomodel.

When a steep gradient is artificially created by painting high-threshold pixels, the segmentation occurs almost exactly along the boundary without this smoothing effect. This results in a somewhat blocky appearance of the tumour in the biomodel.

Other techniques can be used to improve surface quality:

- The tumour may be more easily segmented from an MR dataset and then co-registered to the CT of the bone and vessels.
- The volume-based imaging tools of 'erosion' and 'dilation' where voxel connectivity is used in combination with thresholding to edit the volume.

The close association of the tumours to the temporal bone had both positive and negative effects on the biomodelling process.

A positive effect is that the tumour is supported by the bone during and after manufacture so that additional support structures were unnecessary. If a tumour occurred completely within surrounding tissue that was edited away, it would fall to the bottom of the biomodel or build platform, unless the automatically generated support structures held it in position. It is important that these support structures are attached to other anatomy and not just the build platform, because as the biomodel is removed from the platform the relationship between structures may be destroyed. The anatomy can be rotated so that supports are generated attached to other surfaces, or supports may be manually drawn in using the pixel editor to interconnect the structures, thus maintaining the spatial relationships.

The negative effect is a result of the binary nature of the process at present. When a tumour comes into contact directly with bone it is impossible to see the margin of tumour to bone transition. The same is, of course, true whenever two differing tissue types are modelled together in contact. This problem will be solved when multicoloured resins are made available. One resin can be transparent, allowing the second to be clearly seen through it.

3.2.6   Spinal surgery

The complex anatomy of the spine has been found to be suited to 3D imaging. Improvement in data display by 3D imaging has been reported in traumatic injuries, congenital malformations, degenerative disease and neoplastic conditions (Murphy *et al.*, 1988; Virapongse *et al.*, 1986; Kilcoyne and Mack, 1987; Hadley *et al.*, 1987; Lang *et al.*, 1988; Starshak *et al.*, 1989; Zinreich *et al.*, 1990; Bonnier *et al.*, 1991). The combination of CT/MR scans with plain X-rays when planning spinal surgery yields a complex combination of data. Even with all the standard data, a true appreciation of the anatomy can remain elusive. This is particularly the case with the spine as it is constituted of 28 individual bones, each with complex anatomy and articulation. The overall 3D alignment of the bones over long segments can be particularly difficult to assimilate from standard imaging.

More recently, spinal stereotaxy has been introduced to assist with intraoperative navigation (Kalfas *et al.*, 1995). Spinal stereotaxy is somewhat limited in the spine by the numerous articulations between bones which allow movement to occur in the interval between scanning and surgery or during surgery. Any movement of the spine or individual vertebrae introduces inaccuracy. Trouble with accuracy combined with the complexity of spinal stereotaxy apparatus has limited clinical use.

The author and several colleagues have investigated the utility of biomodelling in the spine (D'Urso *et al.*, 1999e). Studies have been performed on children and adults with congenital, traumatic and degenerative anomalies. It was immediately obvious that the spinal biomodel gave an immediate unique and tactile overview of the anatomy. In all cases a better appreciation of the bony anatomy was demonstrated with the biomodel. In some cases the feasibility of surgery was determined on the basis of the biomodel. Patients also universally stated that the biomodel provided facilitated a better understanding of the anatomy and surgical plan.

3.2.6.1   Spinal biomodel stereotaxy

Commonly, in more complex spinal deformities the placement of pedicle screws and instrumentation is necessary to improve deformity and ensure bony fusion. The placement of such screws can be very difficult in the pathological spine owing to anatomical distortion. In these patients the spinal biomodelling is particularly helpful. The author has developed a technique to assist with the placement of spinal instrumentation, using the biomodel as a stereotactic aid. Preoperatively, a standard electric hand drill is used to drill screw trajectories into the biomodel at the desired locations. As the biomodel is transparent and replicates the thickness of cortical bone accurately, optimal trajectories can be easily and quickly drilled into the biomodel. Metal trajectory pins can then be placed into the drill holes to visualize the trajectory as well as directly to measure the length of the screw required. Once the trajectory pins are *in situ*, cold-cure acrylic cement can be moulded around the pin and the immediate conture of the biomodel to fashion a contour-matching stereotactic drill guide. The template and the biomodel are sterilized and used intraoperatively to navigate and confirm anatomical relationships. The template may then be contour matched to the spine and used as a drill guide to replicate the preoperatively planned trajectories in the patient's spine. A similar technique has previously been described for cranial applications (D'Urso *et al.*, 1999b).

Limitations of the stereotactic template are the time required for its preoperative manufacture and difficulty in contour matching the template intraoperatively. The template must be made so that it does not replicate the immediate anatomy in such complexity that any minor variation would introduce inaccuracy through poor fit. Each template must be fashioned so that it will only contour match a single vertebral body because movement between the bones will again affect the accuracy. A separate template is thus required for each vertebra that requires instrumentation. The manufacture of such templates does require some skill and at least 30 minutes of preoperative surgeon time. Another problem identified by the author is the accuracy of fit of such templates. The complexity of the immediate anatomy can make contour fit difficult, particularly if soft tissue is present. All soft tissue must be removed and the exposure of the anatomy must be sufficient to encompass the entire template area. If attention is paid to these limitations, accurate matching of the template can be obtained, allowing accurate drill trajectories to be achieved.

With experience it has become evident that, by having the biomodel with the trajectory pins sterile in the operative field, screw trajectories can be transferred accurately by visual cue. The entry point of the trajectory can be readily determined by direct comparison of the biomodel anatomy (incorporating the preoperatively determined trajectory) to the intraoperative patient anatomy. Once the entry point is established, the trajectory vector can be easily determined by holding the biomodel with the trajectory pin in the same orientation as the patient and within the direct line of sight of the surgeon. The surgeon can align the pedicle finder to the same trajectory as the pin in the biomodel by direct visual comparison. The length of the screw can easily be determined by direct measure from the biomodel. In the experience of the author, the direct visual alignment method is adequate for almost all cases, and is easy in preparation and execution. Not only does this technique allow for the rapid accurate placement of screws but it also reduces the need for intraoperative X-ray to assist placement. A reduction in X-ray exposure to both surgeon and patient is clearly advantageous for both health and safety and cost reasons.

Biomodels have been found to be particularly helpful for minimal invasive spinal procedures. In such cases surgery is performed via a small incision with the aid of a 'tube-like' retractor. This technique limits the visible anatomy and can make orientation and navigation difficult. The surgeon compensates for this by using intraoperative X-ray. The use of biomodels in minimal invasive procedures has been found to increase confidence and accuracy of screw placement while reducing the use of X-ray.

3.2.6.2   Technical considerations in spinal biomodelling

CT scanning

The most important aspect influencing spinal biomodelling is the anatomy of the spine. The average human spine is around 100 cm long and axial in orientation. As CT scanning is performed axially, the length of spine to be modelled will influence the scanning protocol. If a few vertebrae are all that is required, the scan can be performed in high resolution with 1 mm cuts. If, however, a larger region is required, scanning at high resolution will result in relatively high radiation exposure to the patient. The protocol may have to reduce the slice thickness or in spiral CT the gantry feed pitch to reduce the radiation exposure. If the slice acquired is too thick, interpolation between slices may result in pseudofusion of vertebrae as the disc between has been 'missed' during the scanning. More recently, multislice CT has been introduced. This does significantly reduce radiation exposure while increasing the length of the Z axis that can be scanned. Presently it is recommended that CT scanning be performed on a multislice machine in helical mode.

Layer manufacturing

As the spine is tall and narrow, many layers will be required to build a number of vertebrae in the axial orientation. In SL the manufacturing time is directly related to the number of layers required as each recoating cycle takes a set amount of time between layers. In such situations the spine will take less time to manufacture when built rotated through 90° to minimize the build height. The number of supports should also be taken into consideration, and optimizing the build orientation can reduce build time by ensuring the supports are minimized. These considerations are important when biomodelling any structure with axial length greater than its maximum width, e.g. long bones. We recommend the use of the patented build optimization algorthim within Anatomics™ BioBuild™ software for automatic determination of the optimum build orientation.

3.2.7   Orthopaedic biomodelling

Orthopaedic surgeons have been slow to use biomodels. Other than spinal applications, our experience is limited to 18 cases. This has been surprising as the skeletal anatomy commonly treated by orthopaedic surgeons is ideal for biomodelling. Applications that have been studied include hip anomalies, particularly congenital dysplasia, acetabular trauma and revision arthroplasty procedures. Biomodels have also been used in complex revision surgery for the knee and elbow. As most orthopaedic procedures are planned on the basis of plain X-ray alone, there are likely to be barriers against the introduction of CT scanning purely so that a biomodel can be manufactured. It is likely, however, that a steady increase in the use of multislice CT will lead to an increased use of biomodelling technology.

Another factor that limits the use of biomodelling in orthopaedics is the size of the bones imaged. A pelvis, for example, may easily exceed the size of the build platform of an SLA 250, and must be manufactured in parts that have to be glued together. Such a biomodel is very expensive to make and cumbersome to transport and use. The cost of such large biomodels is a deterrent to their use. A way of reducing this problem is to scale the biomodel down to reduce size. The author has conducted studies to assess the utility of scaling and to evaluate the effect of scaling on biomodel quality and accuracy in comparison with a full-size biomodel. The advantage of reducing the image pixel size (x, y dimensions) by half is that, it reduces the biomodel volume by a factor of 8 and thus the cost of manufacture by a similar proportion. Another advantage is that, for large structures such as the pelvis, a smaller biomodel is easier to transport, handle and store. If the biomodel is being created purely for diagnostic reasons, the reduction in size does not greatly influence its utility so long as it is not too small in comparison with the 0.25 mm steps that are formed during construction. For this reason, and in view of the fact that scaling beyond one-half rapidly becomes less cost effective, it would be advisable keep the biomodel dimensions between 50 and 100%, with a reduction of 25–50% being ideal.

The method that was used to generate the scaled biomodel was straightforward. By scaling the x, y and z dimensions of the pixels within BioBuild™, the biomodel will also be scaled accordingly. Using this technique, it is possible either to reduce or to increase the size of the biomodel as required. Although this technique will be most useful for reducing the size of the biomodel, an increase in size may also hold advantages. This would be most apparent when biomodelling a small intricate structure that is difficult to examine at normal size. An example of this is the internal ear cavity which if generated at double size would be much easier to examine. The cost increase when biomodelling such structures would not be great, as the structures are originally quite small, unless of course they were scaled-up many times. The most important limiting factor on the quality of scaled-up biomodels is again the original data quality and resolution. Any artefacts or stepping in the original data will be more noticeable in a larger biomodel. The extent of the stepping may be reduced by further interpolation between build layers which would be possible as the height and number of layers increase.

3.3 Case Studies

Case 1

A twenty five year old man presented with a subarachnoid haemorrhage. An admission CT scan revealed a cerebellar arterio-venous malformation (AVM). Subsequent angiography revealed a complex AVM with multiple flow related aneurysms on its feeding arteries. Given the patient's young age and the history of haemorrhage, it was decided that surgical excision of the AVM was the safest option. To assist surgical planning, navigation and patient consent for surgery, a biomodel was requested. A helical CT angiogram was performed and the data were prepared using Anatomics™ BioBuild™ software. Stereocol resin was used to allow selective colouration of the vascular structures. The biomodel accurately replicated the AVM (Figure 3.3). Initially the biomodel was used to obtain informed consent from the patient and

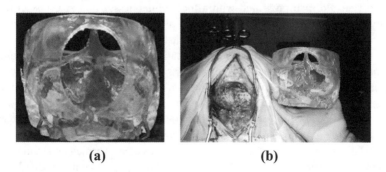

(a)   (b)

Figure   3.3 Case 1: (a) biomodel of a mid-line cerebellar AVM; (b) sterilized biomodel being used intraoperatively

his family. The complexity of the abnormality could easily be appreciated. The risks of the surgery and location of the surgical exposure were also easily explained to the patient and his family. The biomodel was used in conjunction with the preoperative angiogram to plan the surgical approach and the selective obliteration of the feeding vessels to the AVM. The surgeon consulted with an interventional radiologist carefully to select vessels that would be difficult to obliterate surgically. The biomodel was helpful as a communication and planning aid. The surgeon was able to evaluate the access to the AVM and plan the sequence of the surgery. During the operation, the biomodel was invaluable in assisting with surgical navigation and in verifying the location of various vessels supplying the AVM. The biomodel was also used to demonstrate the abnormality to nursing and junior medical staff. Post-operatively the patient made a full recovery and was cured of the AVM.

Case 2

A fifty three year old man presented with seizures caused by a known recurrent parafalcine meningioma. Surgery had been performed on three separate occasions but unfortunately the tumour had recurred. A large part of the skull had been resected and tumour was invading the edges of the cranial deficit as well as the soft tissues and brain. A contrast CT scan was performed. Using Anatomics™ BioBuild™ software, the surgeon marked on the CT scan the limit of tumour invasion with a line. A biomodel was manufactured from StereoCol resin with the line represented by red colouration. The surgeon then used standard equipment to plan and simulate the skull resection on the biomodel. The 'resected' biomodel was then returned to Anatomics™. A custom acrylic implant was then manufactured to reconstruct the defect (Figure 3.4). A custom resection template was also manufactured to allow the surgeon to mark the resection line on the patient's skull. The biomodel, implant and resection template were used to demonstrate the surgical plan to the patient and obtain informed consent. At surgery the patient's scalp was carefully dissected from the tumour and the skull. The resection template was then contour matched to the skull and the resection margin marked. The tumour mass was then resected from the skull and carefully dissected from the brain. The skull was then reconstructed using the custom implant. The patient made a satisfactory recovery, had an excellent cosmetic result and has been tumour free and seizure free since the surgery.

Case 3

A thirty two year old man presented with a focal epileptic seizure. A CT scan revealed an abnormality in his brain suspicious of a tumour. A biopsy was requested. The patient had three non-invasive fiducial markers placed and a contrast CT scan was performed. The data were transferred to an Anatomics™ BioBuild™ workstation and a biomodel was manufactured of the skull, fiducial markers and the tumour with associated blood vessels. The biomodel was used to gain informed consent from the patient. The patient was taken to the operating theatre. The D'Urso stereotactic frame was placed on the biomodel and the trajectory was set and locked into the frame. The biopsy depth was also determined. The frame was then attached to the three fiducial markers and an entry point was marked on the skin. A small linear incision was made and a burrhole was performed. The frame was again verified and then the biomodel was attached to the patient fiducials. The biopsy needle was then gently inserted to the target and biopsies were taken (Figure 3.5). The procedure took 20 min to perform. The biopsy was positive for tumour. The patient made an uneventful post-operative recovery.

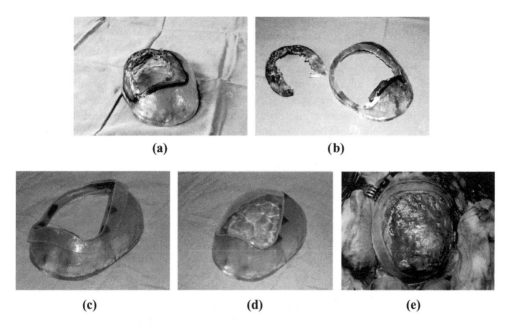

Figure 3.4 Case 2: (a) biomodel with tumour invading skull. Resection margin marked; (b) biomodel with simulated tumour resection; (c) biomodel with resection template; (d) biomodel with custom cranioplastic implant; (e) resection template fitted to patient intraoperatively

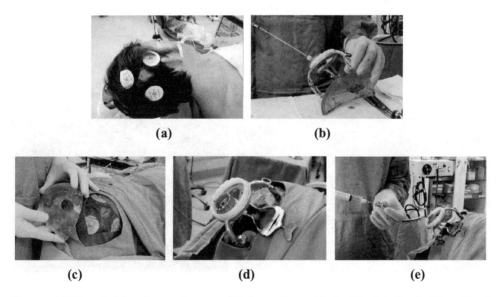

Figure 3.5 Case 3: (a) patient at the start of the procedure with the fiducials fitted; (b) sterilized biomodel next to the patient who is prepared for surgery; (c) stereotactic frame fixed to the biomodel to determine the trajectory; (d) stereotactic frame fitted to the patient; (e) stereotactic biopsy being performed

Case 4

A sixty four year old patient presented with severe low back and leg pain. Previous surgery for a lumbar disc prolapse had been performed on two occasions and had made the patient worse. The patient's imaging revealed a complex lumbar spinal disorder with damaged facet joints, a spondylolthesis (instability) at lumbar 4–5 (L4/5) level and nerve compression. A CT was performed and a biomodel was manufactured. The surgeon used the biomodel to simulate the placement of pedicle screws using trajectory pins and a standard power drill. With the trajectory pins *in situ*, cold-cure bone cement was shaped to create drill guide templates. The biomodel, trajectory pins and custom templates were shown to the patient and her husband. Informed consent was obtained. At surgery, the biomodel was used to navigate and expose the spine while avoiding neural tissue. The biomodel was particularly useful for this as the previous surgery had severely distorted the anatomy. The drill templates were then contour matched to the spine and used to drill the pedicles L4/5. The nerves were decompressed and the spine was brought into alignment. The screws were then used to fix two plates to L4/5 to stabilize the level. Bone graft was harvested and packed around the plates to achieve bony fusion (Figure 3.6). The patient made an uneventful recovery and became virtually pain free.

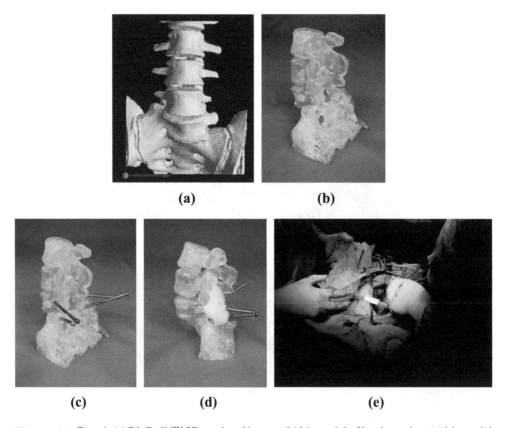

(a)   **(b)**

(c)   **(d)**   **(e)**

Figure   3.6 Case 4: (a) BioBuild™ 3D rendered image; (b) biomodel of lumbar spine; (c) biomodel with stereotactic trajectory pins *in situ*; (d) biomodel with stereotactic guide pins and custom templates; (e) sterilized biomodel being used intraoperatively

References

Abbott, J., Netherway, D., Wingate, P., Abbott, A. and David, D. (1994) Craniofacial imaging, models and prostheses. *Aust. J. Otolaryng.*, **1**, 581–7.

Anichkov, A. D., Polonsky, J. Z., Usov, V. V. (1977) Method of guiding a stereotactic instrument at an intracerebral space target point. US Patent 4,228,799.

Arvier, J. F., Barker, T. M., Yau, Y. Y., D'Urso, P. S., Atkinson, R. L. and Mcdermant, G. R. (1994) Maxillofacial biomodelling. *Br. J. Oral and Maxillofac. Surg.*, **32**, 276–83.

Barker, T. M., Earwaker, W. J. S. and Lisle, D. A. (1994) Accuracy of stereolithographic models of human anatomy. *Australasian Radiology*, **38**.

Binder, T. M., Moertl, D., Mundigler, G., Rehak, G., Franke, M., Delle-Karth, G., Mohl, W., Baumgartner, H. and Maurer, G. (2000) Stereolithographic biomodeling to create tangible hard copies of cardiac structures from echocardiographic data: in vitro and in vivo validation. *J. Am. Coll. Cardiol.*, January, **35** (1), 230–7.

Blake, G. B., MacFarlane, M. R. and Hinton, J. W. (1990) Titanium in reconstructive surgery of the skull and face. *Br. J. Plast. Surg.*, **43**, 528–35.

Bonnier, L., Ayadi, K., Vasdev, A., Crouzet, G. and Raphael, B. (1991) Three-dimensional reconstruction in routine computerized tomography of the skull and spine. *J. Neuroradiol.*, **18**, 250–266.

Brown, R. A. (1979) A computerised tomography-computer graphics approach to stereotactic localization. *J. Neurosurg.*, **50** (6), 715–20.

D'Urso, P. S. (1993) Surgical procedures. Australian Patent PM2398, November.

D'Urso, P. S., Askin, G., Earwaker, W. J. S., Merry, G., Thompson, R. G., Barker, T. M. and Effeney, D. J. (1999e) Spinal biomodelling. *Spine*, **24** (12), 1247–51.

D'Urso, P. S., Atkinson, R. L., Lanigan, M. W., Earwaker, W. J., Bruce, I. J., Holmes, A., Barker, T. M., Effeney, D. J. and Thompson, R. G. (1998) Stereolithographic (SL) biomodelling in craniofacial surgery. *Br. J. Plastic Surg.*, **51** (7), 522–30.

D'Urso, P. S., Atkinson, R. L., Weidmann, M. J., Redmond, M. J., Hall, B. I., Earwaker, W. J. S., Thompson, R. G. and Effeney, D. J. (1999d) Biomodelling of skull base tumours. *J. Clinical Neuroscience*, **6** (1), 31–5.

D'Urso, P. S., Barker, T. M., Arvier, J. F., Earwaker, W. J., Bruce, I. J., Atkinson, R. L., Lanigan, M. W. and Effeney, D. J. (1999a) Stereolithographic (SL) biomodelling in cranio-maxillofacial surgery: a prospective trial. *J. Cranio-maxillofacial Surg.*, February, **27** (1), 30–7.

D'Urso, P. S., Earwaker, W. J., Barker, T. M., Redmond, M. J., Thompson, R. G., Effeney, D. J. and Tomlinson, F. H. (2000) Custom cranioplasty using stereolithography and acrylic. *Br. J. Plastic Surg.*, **53** (3), 200–04.

D'Urso, P. S., Hall, B. I., Atkinson, R. L., Weidmann, M. J. and Redmond, M. J. (1999b) Biomodel guided stereotaxy. *Neurosurgery*, **44** (5), 1084–93.

D'Urso, P. S. and Thompson, R. G. (1998) Fetal biomodelling. *Aust. and New Zealand J. Obstetrics and Gynaecology*, **38**, 205–207.

D'Urso, P. S., Thompson, R. G., Atkinson, R. L., Weidmann, M. J., Redmond, M. J., Hall, B. I., Jeavons, S. J., Benson, M. D. and Earwaker, W. J. S. (1999c) Cerebrovascular biomodelling. *Surgical Neurology*, **52** (5), 490–500.

D'Urso, P. S., Tomlinson, F. H., Earwaker, W. J., Barker, T. M., Atkinson, R. L., Weidmann, M. J., Redmond, M., Hall, B., M'Kirdy, B., Loose, S., Wakely, G., Reik, A. T. and Effeney, D. J. (1994) Stereolithographic (SLA) biomodelling in cranioplastic implant surgery. *Proceedings of the International Conference on Recent Advances in Neurotraumatology*, September 25–28, 1994, Gold Coast, Australia, 153–6.

Eufinger, H., Wehmoller, M., Harders, A. and Heuser, L. (1995) Prefabricated prostheses for the reconstruction of skull defects. *Int. J. Oral and Maxillofac. Surg.*, **24**, 104–110.

Fukuta, K. and Jackson, I. T. (1992) New developments in three-dimensional imaging: clinical application of interactive surgical planning for craniofacial disorders. *Perspect. Plast. Surg.*, **6**, 155–77.

Gillespie. J. E., Adams, J. E. and Isherwood, I. (1987) Three-dimensional computed tomographic reformations of sellar and parasellar lesions. *Neuro-radiol.*, **29**, 30–35.

Gilon, D., Cape, E. G., Handschumacher, M. D., Song, J. K., Solheim, J., VanAuker, M., King, M. E. and Levine, R. A. (2002) Effect of three-dimensional valve shape on the hemodynamics of aortic stenosis:

three-dimensional echocardiographic stereolithography and patient studies. *J. Am. Coll. Cardiol.*, October, **40** (8), 1479–86.

Hadley, M. N., Somtag, V. K. H., Amos, M. R., Hodak, J. A. and Lopez, L. J. (1987) Three-dimensional computed tomography in the diagnosis of vertebral column pathological conditions. *Neurosurgery*, **21**, 186.

Hounsfield, G. N. (1973) Computerised transverse axial scanning (tomography). Part 1: description of system. *Br. J. Radiol.*, **46**, 1016.

Howard, J. D., Elster, A. D. and May, J. S. (1990) Temporal bone: three-dimensional CT. Part II: pathologic alterations. *Radiology*, **177**, 427–30.

Joffe, J. M., McDermott, P. J. C., Linney, A. D., Mosse, C. A. and Harris, M. (1992) Computer-generated titanium cranioplasty: report of a new technique for repairing skull defects. *Br. J. Neurosurg.*, **6**, 343–50.

Kalfas, I. H., Kormos, D. W., Murphy, M. A., McKenzie, R. L., Barnett, G. H., Bell, G. R., Steiner, C. P., Trimble, M. B. and Weisenberger, J. P. (1995) Application of frameless stereotaxy to pedicle screw fixation of the spine. *J. Neurosurg.*, October, **83** (4), 641–7.

Kato, K., Ishiguchi, T., Maruyama, K., Naganawa, S. and Ishigaki, T. (2001) Accuracy of plastic replica of aortic aneurysm using 3D-CT data for transluminal stent-grafting: experimental and clinical evaluation. *J. Comput. Assist. Tomogr.*, March–April, **25** (2), 300–304.

Kilcoyne, R. F. and Mack, L. A. (1987) Computed tomography of spinal fractures. *Appl. Radiol.*, **3** (16), 40–54.

Kobayashi, S., Hara, H., Okudera, H., Takemae, T. and Sugita, K. (1987) Usefulness of ceramic implants in neurosurgery. *Neurosurgery*, **21**, 751–5.

Lang, Ph., Genant, H. K., Chafetz, N., Steiger, P. and Morris, J. M. (1988) Three-dimensional computed tomography and multiplanar reformations in the assessment of pseudarthrosis in postenor lumbar fusion patients. *Spine*, **13**, 69

Lermusiaux, P., Leroux, C., Tasse, J. C., Castellani, L. and Martinez, R. (2001) Aortic aneurysm: construction of a life-size model by rapid prototyping. *Ann. Vasc. Surg.*, March, **15** (2), 131–5.

Manson, P. N., Crawley, W. A. and Hoopes, J. E. (1986) Frontal cranioplasty: risk factors and choice of cranial vault reconstructive material. *Plastic and Reconstr. Surg.*, **77**, 888–900.

Murphy, S. B., Kijewski, P. K., Millis, M. B., Hall, J. E., Simon, S. R. and Chandler, H. P. (1988) The planning of orthopaedic reconstructive surgery using computer-aided simulation and design. *Comput. Med. Imag. Graph.*, **12**, 33–45.

Pentecost, J. O., Sahn, D. J., Thornburg, B. L., Gharib, M., Baptista, A. and Thornburg, K. L. (2001) Graphical and stereolithographic models of the developing human heart lumen. *Comput. Med. Imaging Graph.*, November–December, **25** (6), 459–63.

Poulsen, M., Lindsay, C., Sullivan, T. and D'Urso, P. S. (1999) Stereolithographic modelling as an aid to orbital brachytherapy. *Int. J. Radiation Oncology, Biology and Physics*, June, **44** (3), 731–5.

Ramsden, R. T., Herdman, R. C. D., Lye, R. H. (1992) Ionomeric bone cement in neuro-otological surgery. *J. Laryngol Otol.*, **106** 949–53.

Remsen, K., Lawson, W. and Biller, H. F. (1986) Acrylic frontal cranioplasty. *Head and Neck Surg.*, September–October, 32–41.

Rhodes, M. L., Kuo, Y., Rothman, S. L. G. and Woznick, C. (1987) An application of computer graphics and networks to anatomic model and prosthesis manufacturing. *IEEE CG&A*, **2**, 12–25.

Starshak, R. J., Crawford, C. R., Waisman, R. C. and Sty, J. R. (1989) Three-dimensional CT of the pediatric spine. *Appl. Radiol.*, **18** (11), 15–26.

Toth, B. A., Ellis, D. S. and Stewart, W. B. (1988) Computer-designed prostheses for orbitocranial reconstruction. *Plast. Recon. Surg.*, **81**, 315–24.

Vannier, M. W., Gado, M. H. and Marsh, J. L. (1983) Three-dimensional display of intracranial soft-tissue structures. *Am. J. Neuroradiol.*, **4**, 520–1.

Virapongse, C., Shapiro, M., Gmitro, A. and Sarwar, M. (1986) Three-dimensional computed tomographic reformation of the spine, skull, and brain from axial images. *Neurosurgery*, **8**, 53–8.

Vrielinck, L., Politis, C., Schepers, S., Pauwels, M. and Naert, I. (2003) Image-based planning and clinical validation of zygoma and pterygoid implant placement in patients with severe bone atrophy using customized drill guides. Preliminary results from a prospective clinical follow-up study. *Int. J. Oral Maxillofac. Surg.*, February, **32** (1), 7–14.

Waite, P. D., Morawetz, R. B., Zeiger, H. E. and Pincock, J. L. (1989) Reconstruction of cranial defects with porous hydroxylapatite blocks. *Neurosurgery*, **25**, 214–17.

White, D. N. (1982) Method of forming implantable prostheses for reconstructive surgery. US Patent 4436683, 3 June.

Whitman, D. H. and Connaughton, B. (1999) Model surgery prediction for mandibular midline distraction osteogenesis. *Int. J. Oral Maxillofac. Surg.*, December, **28** (6), 421–3.

Wurm, G., Tomancok, B., Nussbaumer, K., Adelwohrer, C. and Holl, K. (2004) Cerebrovascular stereolithographic biomodeling for aneurysm surgery. Technical note. *J. Neurosurg.*, January, **100** (1), 139–45.

Yamaji, K. E., Gateno, J., Xia, J. J. and Teichgraeber, J. F. (2004) New internal Le Fort I distractor for the treatment of midface hypoplasia. *J. Craniofac. Surg.*, January, **15** (1), 124–7.

Zinreich, S. J., Long, D. M., Davis, R., Quinn, C. B., McAfee, P. C. and Wang, H. (1990) Three-dimensional CT imaging in postsurgical 'failed back' syndrome. *J. Comput. Assist. Tomogr.*, **14**, 574–80.

4

Three-dimensional Data Capture and Processing

W. Feng, Y. F. Zhang, Y. F. Wu and Y. S. Wong

This chapter covers different data capture techniques appropriate to medical applications, from the introduction of the data capture techniques to their applications. It will then go on to discuss the software implications. This is therefore considered to be an introductory chapter for a few of the later chapters.

Many three-dimensional (3D) medical scan technologies can help to collect or capture the initial geometry data. Computed tomography (CT) and magnetic resonance (MR) scan technology is very popular in medical application and the scanned data presentation (layer by layer) strongly resembles the slice data format used to drive the new manufacturing paradigm of rapid prototyping (RP). Encouraged by the similarity, computer translators have been developed to convert the CT/MR data from their virtual reality into a format whereby RP could be used to produce anatomically accurate physical models ('real virtuality'). The precision of RP technology is driving the refinement of the algorithms for identifying surfaces and features from the CT/MR data. The RP models have been found to have particular usefulness in many applications, such as diagnostics, treatment planning, procedure practice/simulation and consultation. Inaccuracies in the RP models result from the limited resolution of the CT/MR data and limitations with current data interpretation algorithms. 3D laser scan technology can give quick results with high resolution, and, based on this advantage, RE and RP can be integrated for medical applications. This chapter describes two methods for reconstructing 3D RP models from the laser-scanned data. Current applications will be illustrated for these physical medical images along with research efforts into generating more accurate surface data.

Advanced Manufacturing Technology for Medical Applications   Edited by I. Gibson
© 2006 John Wiley & Sons, Ltd.

4.1   Introduction

One of the most time-consuming aspects of creating 3D virtual models is the creation of the geometric models of objects. This can be particularly frustrating when there is a real or physical version of the object. Fortunately, there are a variety of commercially available technologies that can be used to digitize objects from the molecular scale up to multistorey buildings. Many of the commercial products are in use at service bureau, so builders with smaller budgets or infrequent demands can avoid the capital and time investments. Many commercial vendors offer sensors, software and/or complete integrated systems.

The process of 3D digitizing basically consists of a sensing phase followed by a reconstruction phase. The sensing phase collects or captures the raw data and generates the initial geometry data, usually as a 2D boundary object, or a 3D point cloud. Sensing technologies are based on tracking, imaging, range finding or their combination. The reconstruction phase is the internal processing of the data into conventional 3D CAD and animation geometry data, such as NURBS and polygon sets. Sophisticated reconstruction software packages are available from scanner vendors and third-party software providers.

Tracking systems digitize by positioning a probe on the object and trigger the computer to record the location. The simplest tracker is a mechanical linkage or pantograph. Coordinate measuring machines (CMMs) are robust 3D mechanical trackers for manufacturing applications. Electromagnetic, ultrasonic, optical, gyroscopic and inertial trackers are also used in some commercial 3D digitizers. Trackers can suffer from interference problems, either mechanical or electromagnetic. Object and environment space and materials need to be considered. Manual tracking systems require a large amount of patient, skilled labour, but they can digitize an object directly into polygonal models, eliminating the need for the reconstruction phase. Automated probe tracking systems produce point cloud data that will require reconstruction. One form of automated tracker is the scanning probe microscope (SPM), which can be used to create 3D models of molecular-scale objects.

Imaging starts with the capture of one or more 2D images that are then used as input to image processing algorithms to create the initial geometry data. Some imaging methods create a point cloud, while others use feature extraction to create an initial topology model. Active imaging systems project a moiré or grid pattern on the objects to provide known reference points and simplify the image processing. Passive imaging systems collect available light. There are some commercial software packages that can use stereoscopic satellite and aerial images to create terrain elevation and building models.

Another image approach to 3D scanning relies on using a series of slices sectioned through the object. These can be obtained by actually cutting the object and taking optical photographs of the slices or by using advanced sensors such as ultrasound, magnetic resonance imaging (MRI), X-ray computed tomography (CT) and confocal microscopes. The slices can be used to produce volumetric data (voxels) or feature extraction might be used on the images to produce contour lines. Both forms of data can be readily converted to polygonal and surface models. Microscopic and volumetric systems are generally very expensive.

Range finding is related to imaging as it usually results in a 2D array of range data (think Z buffer) that is then processed as an image. While 3D Pipeline Corporation (CA, United States) has developed an ultrasonic range finder system to digitize caves and other large structures, optical systems are the dominant range-finding technology. Both laser and white-light range finding are available. Some optical range scanners also capture object colour into an image for use as a texture map.

All imaging technologies require multiple images from different views in order to create a complete object scan. Some scanning systems acquire these by rotating the object, while others combine an imaging (or range-finding) sensor with a tracker device allowing the sensor to be moved around the object. Many of the laser or moiré scanner vendors market their sensors for attachment to CMM mechanical arms.

Not all the 3D digitizing techniques can be used for medical imaging because of the different requirements for medical scanning and industry scanning. This chapter only discusses some medical scanning techniques and their processes.

4.2   3D Medical Scan Process

There are many 3D scan techniques that have been used in medical applications. Computed tomography (CT), magnetic resonance imaging (MRI), ultrasonic imaging and 3D laser scanning are the most commonly used.

The medical scan process consists of a scan phase and a reconstruction phase. The 3D scan phase captures the raw data and generates the initial geometry data, usually as a 2D boundary object or a 3D point cloud. The 3D reconstruction phase is the internal processing of these data into conventional 3D CAD and animation geometry data such as NURBS and polygon sets.

4.2.1   3D scanning

Different scan techniques have different ways of working, and their clinical benefits are different too. They can be used in different medical applications because of their unique features. The following section discusses the most widely used 3D medical scan technologies: CT, MRI, ultrasound imaging and 3D laser scanning.

4.2.1.1   Computed tomography imaging and its applications

Computed tomography (CT) imaging, also known as 'CAT scanning' (Computed Axial Tomography), was developed in the early to mid-1970s and is now available at over 30 000 locations throughout the world. CT is fast, patient friendly and has the unique ability to image a combination of soft tissue, bone and blood vessels. CT is the workhorse imaging system in most busy radiology departments and diagnostic centres. Since its invention some 25 years ago, CT imaging has seen massive advances in technology and clinical performance. Today, CT enables the diagnosis of a wider array of illness and injury than ever before!

CT is based on the X-ray principle: as X-rays pass through the body they are absorbed or attenuated at differing levels, creating a matrix or profile of X-ray beams of different strength. This X-ray profile is registered on film, thus creating an image. In the case of CT, the film is replaced by a banana-shaped detector which measures the X-ray profile.

A CT scanner looks like a big, square doughnut. The patient aperture (opening) is 60–70 cm (24–28″) in diameter. Inside the covers of the CT scanner is a rotating frame which has an X-ray tube mounted on one side and the banana-shaped detector mounted on the opposite side.

A fan beam of X-ray is created as the rotating frame spins the X-ray tube and detector around the patient. Each time the X-ray tube and detector make a 360° rotation, an image or 'slice' is acquired. This 'slice' is collimated (focused) to a thickness of 1–10 mm using lead shutters in front of the X-ray tube and X-ray detector.

As the X-ray tube and detector make this 360° rotation, the detector takes numerous snapshots (called profiles) of the attenuated X-ray beam. Typically, in one 360° lap, about 1000 profiles are sampled. Each profile is subdivided spatially (divided into partitions) by the detectors and fed into about 700 individual channels. Each profile is then backwards reconstructed (or 'back projected') by a dedicated computer into a 2D image of the 'slice' that was scanned.

Multiple computers are used to control the entire CT system. The main computer that orchestrates the operation of the entire system is called the 'host computer'. There is also a dedicated computer that reconstructs the 'raw CT data' into an image. A workstation with a mouse, keyboard and other dedicated controls allows the technologist to control and monitor the exam. The CT gantry and table have multiple microprocessors that control the rotation of the gantry, the movement of the table (up/down and in/out), the tilting of the gantry for angled images and other functions such as turning the X-ray beam on and off.

Unlike other medical imaging techniques, such as conventional X-ray imaging (radiography), CT enables direct imaging and differentiation of soft tissue structures, such as liver, lung tissue and fat. CT is especially useful in searching for large space occupying lesions, tumours and metastasis and can reveal not only their presence but also the size, spatial location and extent of a tumour.

CT imaging of the head and brain can detect tumours, show blood clots and blood vessel defects, show enlarged ventricles (caused by a build-up of cerebrospinal fluid) and image other abnormalities such as those of the nerves or muscles of the eye.

Owing to the short scan times of 500 ms to a few seconds, CT can be used for all anatomic regions, including those susceptible to patient motion and breathing. For example, in the thorax, CT can be used for visualization of nodular structures, infiltrations of fluid, fibrosis (for example from asbestos fibres) and effusions (filling of an air space with fluid).

CT has been the basis for interventional work like CT-guided biopsy and minimally invasive therapy. CT images are also used as the basis for planning radiotherapy cancer treatment. CT is also often used to follow the course of cancer treatment to determine how the tumour is responding to treatment.

CT imaging provides both good soft tissue resolution (contrast) as well as high spatial resolution. This enables the use of CT in orthopaedic medicine and imaging of bony structures including prolapses (protrusion) of vertebral discs, imaging of complex joints like the shoulder or hip as a functional unit and fractures, especially those affecting the spine. The image postprocessing capabilities of CT – like multiplanar reconstructions and 3D display – further enhance the value of CT imaging for surgeons. For instance, 3-D CT is an invaluable tool for surgical reconstruction following facial trauma.

CT is becoming the method of choice for imaging trauma patients. CT exams are fast and simple, enable a quick overview of possibly life-threatening pathology and rapidly enable a dedicated surgical treatment. With the advent of spiral CT, the continuous acquisition of complete CT volumes can be used for the diagnosis of blood vessels with CT angiography. For instance, abdominal aortic aneurysms, the renal arteries, the carotid vessels and the circle of Willis can all now be quickly imaged with CT with minimal intervention.

4.2.1.2 Magnetic resonance imaging and its applications

Magnetic resonance (MR) principles were initially investigated in the 1950s, showing that different materials resonated at different magnetic field strengths. Magnetic resonance imaging (also know as MRI) was initially researched in the early 1970s, and the first MR imaging prototypes were tested on clinical patients in 1980.

MR uses magnetic energy and radio waves to create cross-sectional images or 'slices' of the human body. The main component of most MR systems is a large tube-shaped or cylindrical magnet. Also now available are MR systems with a C-shaped magnet or other type of open design. The strength of the MR systems magnetic field is measured in metric units called tesla. Most of the cylindrical magnets have a strength between 0.5 and 1.5 T and most of the open or C-shaped magnets have a magnetic strength between 0.01 and 0.35 T. A 1.5 T MR system has a magnetic field 30 000 times stronger than the pull of gravity on the Earth's surface.

The patient aperture or bore of the cylindrical magnet is usually 55–65 cm wide (21.6–25.6″), with a total end-to-end length of 160–260 cm (5′ 3″–8′ 6″). Approximately 3% of MR patients suffer from claustrophobia and may not tolerate an MR exam in a traditional cylindrical MR system. These patients may now have the option of having an MR study with an open MR system where typically many sides of the system are open and claustrophobic anxiety is lessened.

To begin the MR examination, the patient is positioned on a special table and positioned inside the MR system opening where the magnetic field is created by the magnet. Each total MR examination typically is comprised of a series of 2–6 sequences, with each sequence lasting between 2 and 15 min. An 'MR sequence' is a data acquisition that yields a specific image orientation and a specific type of image appearance or 'contrast'. Thus, a typical exam can last for a total of 10 min to 1 h, depending on the type of exam being run and the MR system being used.

During the examination, a radio signal is turned on and off, and subsequently the energy that is absorbed by different atoms in the body is echoed or reflected back out of the body. These echoes are continuously measured by the MR scanner and a digital computer reconstructs these echoes into images of the body. The tapping heard during the MR exam is created when 'gradient coils' are switched on and off to measure the MR signal reflecting back out of the patient's body. A benefit of MRI is that it can easily acquire direct views of the body in almost any orientation, while CT scanners typically acquire images perpendicular to the long body axis.

A benefit of MR is that, unlike conventional X-ray or CT imaging, it does not use X-ray radiation. Magnetic resonance imaging is non-invasive and provides exquisite images with excellent contrast detail of soft tissue and anatomic structures like grey and white matter in the brain or small metastatic lesions (cancers) in the liver. In comparison with MR, conventional X-ray provides images of dense structures like bones with good resolution. The X-ray angiogram is the traditional standard for imaging vessels like the carotid arteries in the neck, vessels in the brain, peripheral arm and leg vessels or the coronary arteries which supply blood to the heart. However, conventional angiographic imaging is very labour and time intensive and requires administration of significant amounts of contrast to image the blood vessels. X-ray angiography does not provide good images of the soft tissue organs in the body like the liver or brain.

Like MR, CT also creates detailed cross-sectional images of the body. However, while CT can depict soft tissue structures much better than conventional X-ray, it does not quite have the contrast detail that MR provides. Many diseases, for example certain brain tumours, are more readily apparent in MR images than in the corresponding CT images owing to the better contrast definition of MR.

MR imaging is unique in that it can also create detailed images of blood vessels without the use of contrast media (although there is a trend towards the use of special MR contrast media called Gadolinium when imaging the vessels as well as soft tissue like the brain).

The use of MR imaging as a diagnostic technique continues to grow, allowing the study of more and more body parts. Initially, MR was mainly used to image the brain and spinal column, and each exam could last up to 1 h. However, MR scanners can now image a host of additional body parts including injuries of the joints (such as the shoulder, knee, elbow and wrist), the blood vessels (for instance, carotid arteries, renal arteries and peripheral leg arteries) and the breast, as well as abdominal and pelvic organs like the liver or male and female reproductive anatomy. MR examinations have also become much faster, in some cases rivalling the speed of spiral CT. Skilled MR operators with the latest equipment can now do complete routine studies (e.g. brain or knee) in as little as 10 min. In some cases, MR systems equipped with echo planar imaging (EPI) packages can use special emergency or fast protocols to do a basic head study in as little as 5–20 s.

4.2.1.3 Ultrasound imaging and its applications

In the 1960s the principles of sonar (developed extensively by the US Defense Department during World War II) were applied to medical diagnostic imaging. The ultrasound process involves placing a small device called a transducer against the skin of the patient near the region of interest, for example, against the back to image the kidneys. The ultrasound transducer combines the functions of a stereo loudspeaker and a microphone in one device: it can transmit sound and receive sound. This transducer produces a stream of inaudible, high-frequency sound waves which penetrate into the body and bounce off the organs inside. The transducer detects sound waves as they bounce off or echo back from the internal structures and contours of the organs. Different tissues reflect these sound waves differently, causing a signature that can be measured and transformed into an image. These waves are received by the ultrasound machine and turned into live pictures with the use of computers and reconstruction software.

Ultrasound imaging (also called ultrasound scanning or sonography) is a relatively inexpensive, fast and radiation-free imaging modality. Ultrasound is excellent for non-invasively imaging and diagnosing a number of organs and conditions, without X-ray radiation. Modern obstetric medicine (for guiding pregnancy and child birth) relies heavily on ultrasound to provide detailed images of the fetus and uterus. Ultrasound can show fetal development and bodily function such as breathing, urination and movement. Ultrasound is also extensively used for evaluating the kidneys, liver, pancreas, heart and blood vessels of the neck and abdomen. Ultrasound can also be used to guide fine-needle tissue biopsy to facilitate the sampling of cells from an organ for lab testing (for example, to test for cancerous tissue).

Ultrasound imaging and ultrasound angiography are finding a greater role in the detection, diagnosis and treatment of heart disease, heart attack, acute stroke and vascular disease which can lead to stroke. Ultrasound is also being used more and more to image the breasts and to guide biopsy of breast cancer.

4.2.1.4   3D laser scanning

3D laser scanning is a 3D surface digitizing technology that captures the digital shape of physical objects by using laser triangulation technology. Laser triangulation is an active stereoscopic technique where the distance of the object is computed by means of a directional light source and a video camera. A laser beam is deflected from a mirror onto a scanning object. The object scatters the light, which is then collected by a video camera located at a known triangulation distance from the laser. Using trigonometry, the 3D spatial (*XYZ*) coordinates of a surface point are calculated. The 2D array of the CCD camera captures the surface profile image and digitizes all data points along the laser. Through specialized software, the 3D laser scan data can be easily compared with a CAD file, enabling deviations from normal to be graphically displayed.

The primary advantage of laser scanning is that the process is non-contact and fast and results in coordinate locations that lie directly on the surface of the scanned object. This allows fragile parts to be measured and makes the scanned coordinate locations especially useful to dental or prosthetic applications. The high resolution and thinner beam of the laser also permit scanning of highly detailed objects where mechanical touch probes may be too large to accomplish the task. Also, while many touch probe systems attempt to compute true surface coordinates by sensing probe deflection, there are certain geometries where probe deflection can be 'fooled'.

4.2.2   3D reconstruction

Reconstruction is the abstract 'rebuilding' of something that has been torn apart. In the medical imaging context, it is often necessary to acquire data from methods that essentially 'tear' data apart (or acquire the data one piece at a time) in order to be able to view what is inside. Also, a large part of reconstruction is then being able to view, or visualize, all the data once they have been put back together again. Now this seems to be fairly abstract, but here are some real-world examples:

1. In serial section microscopy, the tissue being studied is sectioned into a number of slices and each slice is put into a microscope. Then, images of each slice are captured. To recreate how the tissue looked before we sectioned it, we must put all the images of all these slices back together again, just as if we were putting the real slices of tissue back together again.
2. In confocal microscopy, the microscope can obtain a single plane of image data without having to slice the tissue. In this case, we do not need to realign the images of the slices much, but just stack them back together and then visualize the result.
3. In CT, the scanner acquires a number of projections, much like an X-ray, from different positions. Then, these different views through the object (or person!) must be 'deconvolved' (meaning combined) to reconstruct the 3D object.
4. In MRI, the imaging device acquires a number of cross-sectional planes of data through the tissue being studied. Since all these planes must be stacked back together to obtain the complete picture of what the tissue was like, MRI entails some amount of reconstruction and a great deal of visualization, too.

It sounds as if putting the slices back together is easier for confocal, MRI or CT, but serial section reconstructions are still needed. The smaller the object to be examined, the more difficult it is to get an imaging technique such as MRI or confocal microscopy to 'see' it, because the information from surrounding areas blurs out the object of interest. Thus, the smallest thing that MRI can 'see' is about $1 \, \text{mm}^3$. For confocal microscopes, the smallest object that is detectable is about $1/10 \, \mu\text{m}$ ($1/10\,000 \, \text{mm}$). However, once the object has been sliced up, other forms of microscopy such as an electron microscope can be used to see objects almost as small as $1/10\,000 \, \mu\text{m}$ ($1/100\,000\,000 \, \text{mm}$). There are even newer 'atomic force' microscopes that even enable detection of individual atoms, but few researchers (as yet) have done 3D reconstructions at this minute level. The problem is, however, that, the smaller the scale, the more artefacts can be introduced, so the reconstruction process becomes much more difficult.

An alternative method to confocal microscopy is the reconstruction of image data via deconvolution (from Kevin Ryan [kryan@cts.com]). Basically, this consists in dividing out the blur introduced by the system (usually a microscope). The data, d, are produced by the object, o, modulated through (convolved by) the optical transmission function of the imaging system, the point spread function, p. This is standard linear shift invariant filtering, resulting in

$$d = o \, (^*) \, p$$

where $(^*)$ is a convolution.

In the frequency domain, given the Fourier transforms of these items, the equation is

$$D = O \, ^* \, P$$

where * is a multiplication.

Deconvolution is the inverse operation:

$$O = D \, ^* \, P^{-1}$$

In microscopy, where many of these techniques are implemented, it should be noted that the blur in Z is much greater than the blur in X and Y. Also note that this is an ill-conditioned problem: the inverse operation results in dividing by zero in many places. There are various signal processing approaches to this problem, but the end result is a noise-limited approximation to the original object, with some spatial frequencies missing.

Full deconvolution is computationally expensive, but there are approximations that give decent results. The most common is 'serial plane deconvolution', where the images $\pm$ in Z from the plane of interest are used to approximate the blur in the centre plane, and the approximate blur is subtracted. Full deconvolution, for more accuracy, is usually done in an iterative decent fashion, homing in on an answer O' that, when convolved with P, results in a close approximation of the data D. Full deconvolution is accurate, but, as this involves 10–50 iterations of 3D Fourier transforms, it is slow. There is only one commercial product with full deconvolution of which the author is currently aware. Serial plane (or adjacent plane) deconvolution is less accurate in its reconstruction but computationally cheaper – four 2D fast Fourier transforms (FFTs) per plane for the first plane, and two 2D FFTs per plane for each additional plane.

Deconvolution allows full-field imaging with a standard camera, provided care is taken to account for uneven illumination and background bias. Deconvolution is better than confocal acquisition with small point objects (as they resemble the noise spectra of confocal acquisition), while it is weaker on large structureless objects (as it is dependent on image data to calculate the object, and structureless objects give less to work with).

Cryoelectron microscopy (virus reconstruction) (from Stephan Spencer [sspencer@rhino .bocklabs.wisc.edu]) in combination with image reconstruction can yield a 3D structure that includes both surface and internal features and contains a large degree of structural detail (currently at resolutions as low as approximately 20 Å). Image reconstruction is a computationally demanding process based on the assumption that each virus image from the raw data – the 2D electron micrographs – is a 2D projection of the same 3D object. The reconstruction process takes advantage of the icosahedral symmetry of viruses in assigning an orientation to each projection. The projections are subsequently reconstructed into a 3D array of variable electron density values. The array is then rendered using visualization techniques. Because of the assumption of icosahedral symmetry, icosahedrally symmetrical structures are reinforced and any non-icosahedrally symmetrical (e.g. flexible) structures present on or within the particle are averaged out. The gain in overall detail is marked, despite a lack of detail in non-icosahedrally symmetrical structures.

4.3   RE and RP in Medical Application

Medical application of reverse engineering is very necessary for the following reasons:

- the digital model does not exist;
- the shapes of medical objects are very complex.

As shown in Figure 4.1, the goal of RE and RP in medical application can be stated as follows: given data points X obtained from an unknown surface (US), construct an RP model (R) that approximates the US, in the sense of shape error controlling.

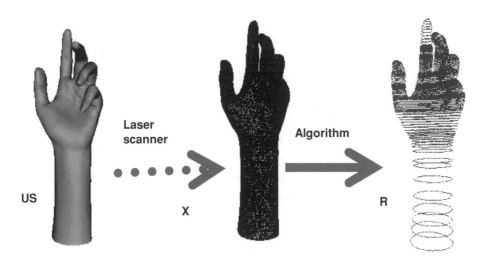

Figure   **4.1** RE and RP in medical application

4.3.1   Proposed method for RP model construction from scanned data

As shown in Figure 4.2, the process of RE starts with a data acquisition phase where raw measurement data are collected from the physical object. Some preprocessing on this data is required in order to combine multiple measurements from different viewpoints. The most crucial phase of the process is the transformation from the 3D cloud data into CAD representation.

The approaches in this phase can be classified into two categories (Wu *et al.*, 2004): the triangular polyhedral mesh based method and the segment and fit based method. The advantages and disadvantages of these two methods are also discussed by Wu *et al.* (2004).

To generate physical objects from CAD models directly, rapid prototyping can produce the physical part by adding layer by layer. Rapid prototyping (RP) is an emerging, non-traditional fabrication method and has been recognized as a valid tool to shorten the lead time from design to manufacture effectively. A variety of RP technologies have emerged (Yan and Gu, 1996). They include stereolithography (SLA), selective laser sintering (SLS), fused deposition manufacturing (FDM), laminated object manufacturing (LOM) and three-dimensional printing (3D printing). Among these technologies, the advantages and disadvantages were discussed by Chua and Leong (1996).

In general, modelling point clouds for RP can be realized by three different approaches (Wu *et al.*, 2004). As shown in Figure 4.2b, in the first approach, a surface model is reconstructed from the point cloud and is closed up as a solid. Then, this solid can be sliced on the basis of its geometry information or can be converted to an RP file format, such as STL, which will be sliced by commercial software. The second approach creates an STL-format file of a model directly from the point cloud (e.g. triangulation), and this STL file can be fed into the RP machine directly. The RP machine can slice the STL model. The third approach goes directly from point cloud to an RP slice file (layer-based model). This slice file need not be further sliced by RP machine.

As seen in the previous section, the surface model generated from the first approach has the advantage that it can be edited. However, the shape error of the final RP model (between the RP model and the cloud data) comes from three sources:

- the shape error between the cloud data and the surface model;
- the shape error between the surface model and the STL model;
- the shape error between the STL model and the layer-based RP model.

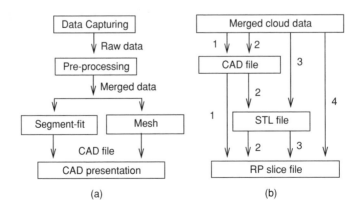

(a)   (b)

Figure   **4.2** RE phases and point cloud data modelling for RP fabrication

This will make the shape error of the RP model very difficult to control. The model generated from the second approach is effectively an STL model. The shape error of the final RP model comes from two sources:

- the shape error between the cloud data and the STL model;
- the shape error between the STL model and the layer-based RP model.

Still, the control of the final shape error is not straightforward. In the third approach, a layer-based model is directly generated from the cloud data, which is very close to the final RP model. Therefore, there is only one source of shape error. If this error can be controlled effectively, this approach will have a clear advantage over the other two modelling approaches in terms of shape error control on the RP model.

We have developed two methods for adaptive slicing. One is adaptive neighbourhood search (ANS) based adaptive slicing which uses a correlation coefficient to determine the neighbourhood size of projected data points, so that we can construct a polygon to approximate the profile of projected data points. It consists of the following steps:

1. The cloud data are segmented into several layers along the RP building direction.
2. Points within each layer are treated as planar data and a polygon is constructed for best fitting of the points.
3. The thickness of each layer is determined adaptively such that the surface error is kept within a given error bound.

The other method is wavelet-based adaptive slicing which uses wavelets to construct a polygon. The general steps are similar to those of the first method. However, the most important step, polygon construction, is different. This has two main stages: Firstly, the near-maximum allowable thickness for each layer is determined with control of the bandwidth of projected points. This estimated bandwidth is controlled by user-specified shape tolerance. Secondly, for each layer, the profile curve is generated by the wavelet method. The boundary points between two regions in one layer are extracted and sorted by a tangent vector based method, which uses a fixed neighbourhood size to quicken the sorting process. The wavelets are then applied to the curve construction from sorted data points from coarser to finer level under the control of the shape error.

The wavelet-based method has better error control and is more robust than the first method because fewer parameters are used. Moreover, the approach is fast since a fixed neighbourhood size is applied in the sorting process, fast wavelet decomposition and reconstruction are used for curve construction and a parallel algorithm can be used for curve construction in different layers.

In our work, the algorithms can deal with cloud data and model construction where:

1. The cloud data comprise an unorganized, noisy sample of an unknown object.
2. This unknown object (surface) can be of arbitrary topological type.
3. No other information, such as structure in the data or orientation information, is provided.
4. The constructed RP model has the same shape errors as the real product, if we ignore the machine error of the RP machine.

4.3.2   Reconstruction software

Many image reconstruction software packages specifically target the medical market and few support the file formats recognized by rapid prototyping systems or other equipment, such as computer numerically controlled (CNC) machines. Without software specifically designed to address the transformation of volumetric data to complex design and 3D rendering, the majority of image reconstruction packages will not be able to deliver what is needed for reverse engineering applications. There are some commercialized imaging software packages (e.g. MIMICS from Materiliase and BioBuild from Anatomics) which can create computer reconstructions of 3D objects taken directly from CT/MRI datasets. Most of these commercialized software packages have rapid prototyping interfaces. Detailed information about medical imaging processing software can be found in Chapter 3 of this book.

4.3.3   Accuracy issues

A medical scan is now a fairly routine procedure. However, a number of criteria must be borne in mind to ensure the acquisition of useful data. It is very important to have a good working relationship with hospital personnel, such as the responsible radiologist, radiology technicians and, if necessary, the patient's physician. Many companies involved in medical RP use a standard form that they supply to the radiologist and radiology technicians, detailing the results desired. The following list presents information and/or parameters that must be known before working with patient scan data:

1. Type of scanner. Determine the model and make of the CT or MR machine. Then check to make sure that the image reconstruction software can translate the data.
2. Kind of scan: axial or helical.
3. Slice thickness: 1.0 mm recommended.
4. Scan spacing: 0.5 mm or at least half the smallest dimension of interest.
5. X-ray strength in the case of CT, pulse sequence in the case of MR.
6. Resolution. Options include image dimensions of 256×256, 512×512 and, less commonly, 1024×1024 pixels. Pixels are stored as 8, 16 or, on some machines, 32 bit of grey-level information per pixel.
7. Field of view (FOV). The object imaged should fill the field of view without extending beyond it.
8. $X-Y$ dimensions of a single pixel. These dimensions and the scan spacing determine the resolution of the coordinate system for reconstruction.
9. Position. The long axis of the object should be parallel to the bore of the scanner. Generally, scans should start just off the object and finish off the other side of the object (so that the entire object is imaged). Objects to be scanned should not be placed on similarly dense objects that will show up in the scan.
10. Artefacts. If significant variations in material densities exist within the object to be scanned, distortion can be experienced. In the case of metal artefacts, the distortion can be severe. The scan protocol can and should be adjusted to take into account the presence of artefacts.
11. Slice time: 2 s/slice is recommended.

Images reconstructed at 512×512 pixels with 16 bit/pixel resolution should require about 0.5 Mbytes of memory per slice. Average datasets can be expected to range from 25 to 100 Mbytes. From the image data, the reconstruction software is then used to extract part contours and/or surfaces, as the case may be. Many thousands of internal and external measurements are quickly generated from the data. Depending on the amount of data and the performance of the software, processing takes from seconds to minutes on a PC. There is no inherent difference between inside and outside, hidden or visible. All features in the object are present in the image data and can therefore be extracted by the software with no penalty in scan time. Moreover, defects are captured as well. If they are important, they can also be extracted and characterized. If only an ideal description of the part is important, defect information can be discarded. The end result is a 3D model that should be exportable, in different file formats, to allow interfacing with other design environments.

A major consideration when using hospital scanners is that the part might not be scanned immediately owing to patient load requirements. Medical technicians may also insist that you provide specific scanning protocols written for your particular requirements. With hospital CT scanning systems the largest matrix can typically be expected to be 512×512 pixels. The lowest FOV would be 9.6 cm, which would result in an X–Y pixel dimension of 0.19 mm $\times$ 0.19 mm (where the Z resolution is 0.5 mm).

With good technique and data, CT scan accuracy generally falls within $\pm 20\%$ of the slice data. For a 1 mm slice this would be equal to ± 0.2 mm. The accuracy of the reconstruction can be influenced by the skill of the image reconstruction operator and the strength of the mathematical algorithms within the reconstruction software. Accuracy of the re-engineered components will also be influenced by the rapid prototyping or tooling technique used to produce the physical representations.

The slice or scan spacing is critical for 3D model reconstructions and should not be confused with slice thickness. Anything over 3 mm is generally not acceptable for complex structures. Slice spacing determines spatial accuracy. The accuracy in the Z axis is determined by the spacing and, if possible, should be at least half the size of the smallest feature that is to be reconstructed.

Lastly, medical reconstruction requires a good understanding of anatomy, which can only come with experience, and an understanding of the types of tissue that are preferentially imaged by CT and MR scanners.

4.4 Applications of Medical Imaging

The physical realization of the virtual data represented by the CT/MR scans has made a significant impact in several areas:

- communication: doctor–patient and doctor–doctor;
- visualization: diagnostic, treatment planning;
- simulation: optimization, stereotactic guides;
- prostheses: implants, plates, etc.

It can also be further used to generate STL or CLI files for RP processing to fabricate 3D physical RP models.

The physical model simplifies communication: both doctor–patient and doctor–doctor. The models are useful for diagnostics and treatment planning, particularly for complex surgical procedures. The accuracy and material properties of the model provide the basis for simulation and evaluation of alternative scenarios, so that optimization can be made prior to the actual surgery. The 'best form' results can be used as templates and for stereotactic feedback to guide the actual surgery. Also, the models can be used to develop accurate prostheses and surgical aids, in advance.

4.5   Case Study

4.5.1   Case study with CT/MR scanned data

The following case depicts a typical medical scan process from scan phase to 3D data reconstruction phase. A skull of a patient was chosen. Part of the patient's skull was removed by surgery. A biocompatible tissue part needed to be produced to fill in the hole and help the patient to heal the region again. The image processing segmentation was performed using tools provided by the Materialise software package MIMCS and CTM module. Mimics is a fully integrated, user-friendly 3D image processing and editing software based on CT or MRI data. The software imports scanner data in a wide variety of formats and offers extended visualization and segmentation functions. Mimics translates CT or MRI data into full 3D CAD, finite element meshes or rapid prototyping data within minutes. By the separation into different modules it can be tuned for every situation.

Step 1. Medical imaging scan
A 3D medical imaging scan is performed on a patient to capture the data needed for the construction of medical imaging processing software. MIMICS of Materialise is used in this case study.

Step 2. Data preparation
The data from the medical imaging scan are prepared using MIMICS software. Data can be sent by tape, optical disk or internet. Data conversion and 3D reconstruction is performed, and the tissue structures of interest are defined, optimized and prepared for construction. Before performing segmentation and visualization, the scan data are imported and converted to MIMICS data format (Figure 4.3).

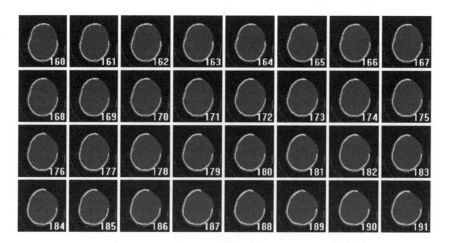

Figure   **4.3** A skull's CT scan images imported in MIMICS

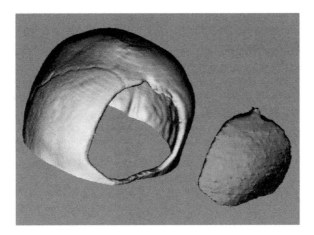

Figure 4.4 3D reconstruction and implant design in MIMICS

Step 3. 3D reconstruction

Images are processed by using a threshold value to differentiate regions of interest. The bone can be separated from soft tissue by setting a fixed threshold value. MIMICS can generate and display a 3D model of the skull (Figure 4.4). There is a hole on the side of the skull. By using MIMICS editing tools and segmentation tools, the missing (hole) part of the skull can be produced and then exported as an STL file. Other software can also be used for similar implant design. Then the STL file generated by MIMICS can be used by a RP system to produce a 3D medical model.

Step 4. RP modelling

There are many commercially available RP systems, and some of them are suitable for medical applications, especially SLA and FDM systems. In this case study both systems are used to produce the same skull 3D model exported from MIMICS (Figure 4.5). The implant designed has been fabricated too.

Figure 4.5 RP models fabricated by FDM (left) and SLA systems

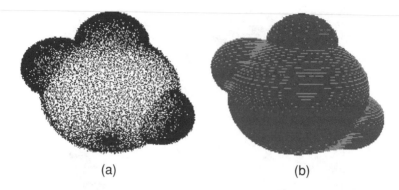

(a)   (b)

Figure   4.6 Original cloud data and the direct RP model of the first case study

4.5.2   Case studies for RE and RP

The algorithms for direct RP model construction based on two methods, ANSAS and WAS, have been implemented with C/C++ in the OpenGL environment. Both simulation results and real case studies are presented here to illustrate the efficacy of the algorithm. The simulation case studies are based on simulated datasets in which the original cloud data are generated by mathematical equations, so that the theoretical shape errors can be obtained accurately and comparison can be made directly. The real case is based on measured data points obtained by a laser scanner, and the results after processing are input to a RP machine for fabrication.

The first case study uses an object composed of four spherical patches (see Figure 4.6a), and it was sliced by the ANS method. As shown in Figure 4.6b, the RP model has 88 layers with a shape tolerance of 0.06.

The second case study is that of a toy cow, as shown in Figure 4.7. The original object can be boxed in a volume of 150 mm × 120 mm × 90 mm and was digitized by a laser scanner – a Minolta VIVID-900 digitizer. The datasets were obtained from different view angles to produce a cloud dataset of 1 098 753 points. The adaptive slicing algorithm was applied to the cloud data employing an error tolerance of 0.7 mm and an initial neighbourhood radius of 0.2 mm. This resulted in a direct RP model, as shown Figure 4.7b with 115 layers.

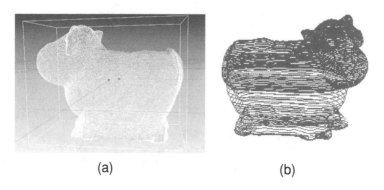

(a)   (b)

Figure   4.7 Original cloud data and the direct RP model of the second case study

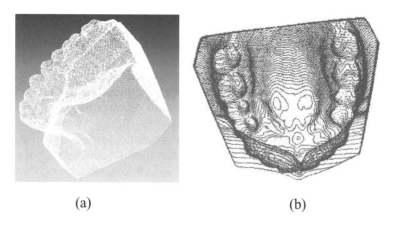

(a) (b)

Figure 4.8 Original cloud data and the direct RP model of the third case study

The part used in the third case study is a lower jaw model, as shown in Figure 4.8. The original object can be boxed in a volume of 78 mm × 72 mm × 52 mm and was digitized by the laser scanner (Minolta VIVID-900 digitizer). The datasets were obtained from different view angles to produce a cloud dataset of 276 591 points. The wavelet-based adaptive slicing algorithm was applied to the cloud data employing an error tolerance of 0.8 mm, with an initial layer thickness of 0.2 mm. This resulted in a direct RP model as shown Figure 4.8b with 68 layers.

The fourth case study uses a prosthetic object and the data points were obtained by a laser scanner (see Figure 4.9a), and it was sliced by the wavelet-based method. As shown in Figure 4.9b, the RP model has 29 layers with a shape tolerance of 3 mm.

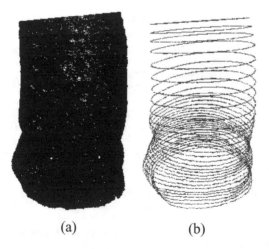

(a) (b)

Figure 4.9 Original cloud data and the direct RP model of the fourth case study

4.6   Conclusions

This chapter discusses a variety of medical imaging scan procedures and the data processing with medical imaging process software. Processing medical scan images and integrating the layered information into a 3D structure can generate accurate virtual anatomical models. The models can be converted into physical realities using rapid prototyping. The resulting models are particularly useful for communications, visualization and simulation, prior to fabrication of prostheses and many other applications. The integration of RE and RP provides fast reconstruction of layered RP models from cloud data with shape error control. In surgical procedures, these models can provide stereotactic feedback, practice simulators, procedure guides (templates), models against which to measure progress and a host of other uses. While the overall dimensional precision of the models is quite high, some technical issues affecting accuracy have been identified, and these are being addressed.

References

Chua, C. K. and Leong, K. F. (1996) *Rapid Prototyping: Principles and Applications in Manufacturing*, John Wiley & Sons, 22–83.

Wu, Y. F., Wong, Y. S., Loh, H. T. and Zhang, Y. F. (2003) Modelling cloud data using an adaptive slicing approach. *Computer Aided Design*, **36** (4), 231–40.

Yan, X. and Gu, P. (1996) A review of rapid prototyping technologies and systems. *Computer Aided Design*, **28** (4), 307–318.

Bibliography

Chelule, K. L., Coole, T. J. and Cheshire, D. G. (2000) Fabrication of medical models from scan data via rapid prototyping techniques. Proceedings of TCT (Time-Compression Technologies) 2000 Conference and Exhibition, October 2000, Cardiff, UK, 45–50.

Feng, W., Wong, Y. S., Fuh, J. Y. H., Hutmacher, D. W. and Teoh, S. H. The application of rapid prototyping robotic dispensing (RPBOD) system for tissue engineering. Proceedings of the 30th International Conference of Computers and Industrial Engineering, June 29–July 2, 2002, Greece, 233–8.

Feng, W., Wong, Y. S. and Hutmacher, D. W. (2004) The application of image processing software for tissue engineering. Proceedings of Asia Pacific Biomech 2004, March 25–28, 2004, Japan.

Griffin, A., McMillin, S. and Knox, C. (1998) Practical examples: using scan data for reverse engineering. SEM technical paper PE n PE98-1221998, 9 pp.

Haystead, J. (1997) Computed-tomography-based medical imaging. *Visions Systems*, July, **2** (7), 14–19.

Hosni, Y. (1997) Advances in rapid prototyping and reverse engineering. Workshop, the 22nd International Conference on Computers and Industrial Engineering, December 1997, Cairo, Egypt.

Hosni, Y., Nayfeh, J. and Harrysson, O. (1999) *Advances in the Medical Applications of Rapid Prototyping*. Proceedings of 8th IIE Research Conference (Research 99), May 1999, Phoenix, AZ, IIE Press.

Imaginis the Breast Health Specialists Medical Procedures Pages: http://imaginis.com/procedures/

Jacobs, P. F. (1992) *Rapid Prototyping and Manufacturing*, Society of Manufacturing Engineers, 1992, ISBN 0-87263-425-6.

Kai, C. C., Jacob, G. G. K. and Mei, T. (1997) Interface between CAD and rapid prototyping systems. Part 1: a study of existing interfaces. *International Journal of Advanced Manufacturing Technology*, **13**, 566–570.

Lightman, A. (1998) Image realization – physical anatomical models from scan data. SPIE Symposium, Medical Imaging, 1998, San Diego, CA.

Materialise (2004) *MIMICS User's Manual*, Materialise, Belgium, April.

Mcgurk, M., Aimis, A. A., Potamianos, P. and Goodger, N. M. (1997) Rapid prototyping techniques for anatomical modelling in medicine. *Ann. Royal Coll. Surgery Engl.*, **79**, 167–74.

Pommert, J. K. *et al.* (1995) Three dimensional imaging in medicine: method and applications, in *Computer Integrated Surgery: technology and clinical applications* (eds R. H. Taylor *et al.*), MIT Press, Cambridge, MA, Ch. 9, 155–74.

Stanford University Medical Center Reconstruction website: http://biocomp.stanford.edu/3dreconstruction/

Udupa, J. K. and Goncalves, R. J. (1995) Imaging transforms for volume visualisation, in *Computer Integrated Surgery: technology and clinical applications* (eds R. H. Taylor *et al.*), MIT Press, Cambridge, MA, Ch. 3, 33–57.

Waterman, P. (1997) 3D Digitizing. *Desktop Engineering*, August, **2** (12), 51–7.

5

Software for Medical Data Transfer

Ellen Dhoore

5.1 Introduction

Thorough preoperative planning of complex surgery is a prerequisite for a successful treatment outcome. Presurgical planning of hard tissue is based mainly upon conventional CT images or MRI images of the patient.

Two different approaches can be defined to efficient 3D presurgical planning of hard-tissue surgery. A first approach is planning on a physical model of the bones. Physical models are based upon CT information and produced by means of rapid prototyping techniques. Surgical actions can be rehearsed on the physical models. In the case of interventions requiring custom implants, implants can be designed on the physical model. An alternative for this approach is working directly on the computer model.

Both approaches require preparatory handling of the CT and/or MRI images by specialized medical imaging software which will be discussed in detail in this chapter.

Dental implantology is one field where preoperative planning is extremely important because of anatomical as well as aesthetic considerations. As an example, this chapter will also elaborate on a dental planning software and existing links to transfer the virtual planning to real surgery.

5.2 Medical Imaging: from Medical Scanner to 3D Model

5.2.1 Introduction

A strong software interface is needed to accept input from every type or brand of scanner, to facilitate the selection of an anatomic structure visible in the medical images and to link this

Advanced Manufacturing Technology for Medical Applications   Edited by I. Gibson
© 2006 John Wiley & Sons, Ltd.

information to rapid prototyping machines for actual model production. Surgery simulation on a virtual computer model requires specific functions in the software to simulate surgical actions and to calculate certain parameters such as volume distance and bone density. Besides pure simulation of surgery actions, links to other software packages (CAD, FEA, CFD, etc.) can facilitate or even be necessary to complete the surgery simulation.

5.2.2   Mimics®

Materialise NV (Leuven, Belgium) provides high-quality solutions supporting clinicians in diagnosis and decision-making. In the digital age, computers and information technology have become a critical factor in reducing costs and improving efficiency in medical environments. Materialise is the worldwide leader in rapid prototyping modelling technology and provides surgeons with the most detailed and precise virtual 3D models available. Clinicians around the world use rapid prototyping models created by Materialise software to assist them in the most complex surgical cases.

Materialise's interactive medical image control system (Mimics) is an interactive tool for the visualization and segmentation of CT images as well as MRI images and 3D rendering of objects. Therefore, in the medical field, Mimics can be used for diagnostic, operation planning or rehearsal purposes. A very flexible interface to rapid prototyping systems is included for building distinctive segmentation objects.

The software enables the surgeon or the radiologist to control and correct the segmentation of CT and MRI scans. For instance, image artefacts coming from metal implants can easily be removed. The object(s) to be visualized and/or produced can be defined exactly by medical staff.

Additional modules provide the interface towards rapid prototyping using STL or direct layer formats with support. Alternatively, an interface to CAD (design of custom-made prostheses and new product lines based on image data) or to finite element meshes is available.

5.2.2.1   Basic functionality of Mimics

Mimics displays the image data in several ways, each providing unique information. Mimics divides the screen into three or four views: the original axial view of the image, and resliced data making up the coronal and sagittal views and 3D view (Figure 5.1). Several visualization functions are included such as contrast enhancement, panning, zooming and rotating of calculated 3D images. Colour scales are used to enhance small differences in the soft tissue or the bone. The alignment or scout image can be displayed. The original data can be resliced online or a resliced project can be exported. Online reslicing allows the display of cross-sectional and parallel images that are orthogonal and along a user-drawn curve in the axial view. The export resliced project tool offers an interface that allows the export of resliced Mimics projects along a user-drawn line. This line can be drawn in any view and in any direction.

Mimics enables different types of measurement to be performed. Point-to-point measurements are possible on both the 2D slices and the 3D reconstructions. A profile line displays an intensity profile of the grey values along a user-defined line. Accurate measurements are possible on the basis of the grey values using three methods: the threshold method, the four-point method and the four-interval method. These methods are ideal for technical CT users. Density

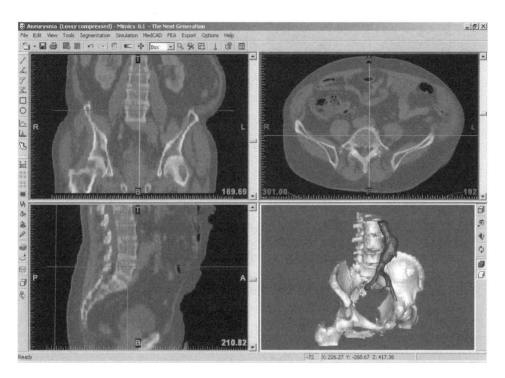

Figure 5.1 Overview Mimics layout

measurements can be performed in ellipses or rectangles: area, mean, grey value and standard deviation are displayed.

Segmentation masks are used to highlight regions of interest. Mimics enables you to define and process images with several different segmentation masks. To create and modify these masks, the following functions are used:

• Thresholding is the first action performed to create a segmentation mask. You can select a region of interest by defining a range of grey values. The boundaries of that range are the lower and upper threshold value. All pixels with a grey value in that range will be highlighted in a mask.
• Region growing will eliminate noise and separate structures that are not connected.
• Editing (draw, erase, local threshold): manual editing functions make it possible to draw, erase or restore parts of images with a local threshold value. Editing is typically used for eliminating artifacts and separating structures.
• Dynamic region growing segments an object on the basis of the connectivity of grey values in a certain grey value range. It allows for easy segmentation of tendons and nerves in CT images, as well as providing an overall useful tool for working with MRI images.
• Morphology operations act on the 'form' of a segmentation mask. All these functions remove or add pixels from the source mask and copy the results to a target mask. This tool is extremely effective when working with MRI images.

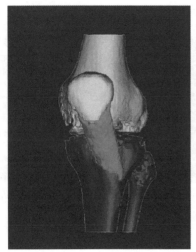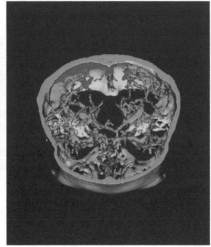

Figure **5.2** Examples of 3D objects in Mimics

- Boolean operations enable different combinations of two segmentation masks (subtraction, union and intersection) to be made. These operations are very useful for reducing the work needed to separate joints.
- Cavity fill fills the internal gaps of a selected mask and copies the result to a new mask. The filling process can be applied in 2D.
- Cavity fill from polylines creates a segmentation mask, starting from a polyline set. This tool is very useful for filling internal cavities in preparation of files for FEA.

After isolation of the region of interest, a 3D can be calculated. Therefore, parameters for resolution and filtering can be set. Information about height, width, volume, surface, etc., is available for every 3D model. Mimics can display the 3D model in any of the windows with visualization functions that include real-time rotation, pan, zoom and transparency. The ability to apply advanced rendering with OpenGL hardware acceleration offers high-quality rendering including Gouraud shading for optimal display of the 3D objects (Figure 5.2).

5.2.2.2   Additional modules in Mimics

Mimics consists of different modules. Figure 5.3 shows the links between the main program and its modules. The different modules link Mimics to various application fields: rapid prototyping (RP), finite element analysis (FEA) and computer-aided design (CAD).

Import module
Mimics imports 2D stacked uncompressed images such as CT, MRI or microscopy data in a wide variety of formats, as well as offering a user-defined import tool. The import software provides direct access to images written on proprietary optical disks and tapes, converts them into the Materialise image format and preserves all necessary information for further processing.

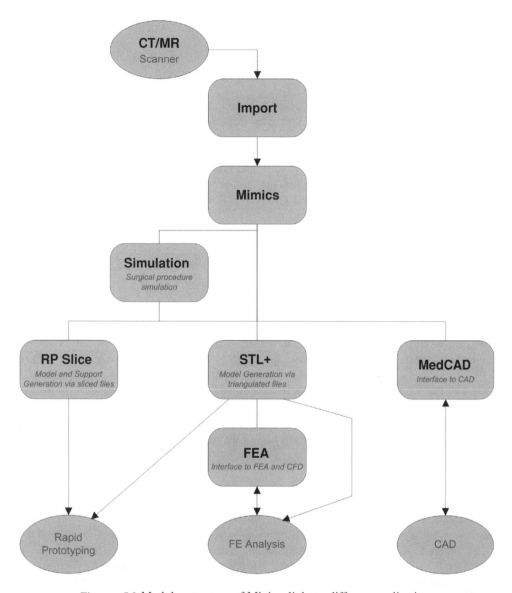

Figure 5.3 Modular structure of Minics: links to different applications

A wizard helps to guide you through the import process. It allows for the merging of multiple image sequences into one project, to convert different image sets at once or to select specific images prior to creating the project.

RP slice module
The RP slice module interfaces from Mimics to any kind of rapid prototyping system via sliced formats and performs support generation.

When creating sliced files, a bilinear and interplane interpolation algorithm is used to enhance the resolution of the RP model. RP slice achieves optimal accuracy in a very short time by the direct conversion of the images to several sliced machine file formats: SLI and SLC for 3D Systems and CLI for Eos. High-order interpolation algorithms result in excellent surface reproduction from scan to model.

RP slice supports colour stereolithography: tumours, teeth and teeth roots and nerve channels can be highlighted in the RP model, giving an extra dimension. Patient information can be displayed by punched or coloured label.

One of the major difficulties in stereolithography and most other layer manufacturing techniques is the need for support structures. Both the generation (automatic or manual) and the removal (cleaning work) of these support structures are complex problems.

The basic function of the support structure is to support the part during the building process. The whole part is connected to a platform, and 'islands' that are isolated at a certain moment during the process need to be attached to the rest of the part. Another function of the supports is to reduce curling effects. Stereolithography resins have a tendency to deform during the building process because of internal stresses generated by the shrinkage. By building a strong support structure under a part, this deformation can be minimized.

When stereolithography is used starting from a CAD representation, the support structures can be designed in the CAD system, which is a large operation. In some instances, however, it is not possible to generate a support starting from an STL description. This is the case in medical applications where the layer information of the CT scanner is interfaced directly to the layer information of the stereolithography machine. This means that there is no surface information available and the standard techniques for automatic support generation cannot be used. In addition, the manual generation of the supports is impossible because the information is not present in a CAD system.

The RP slice module calculates automatically the support structures for the sliced file formats (SLI, SLC and CLI) needed in the production of the rapid prototyping model, starting from contour files.

Case study of RP models: distraction osteogenisis and mandibular reconstruction in a patient with Goldenhar syndrome
Introduction
Hemifacial microsomia is a syndrome in which an underdevelopment exists of one side of the face compared with the other side. In some children only an ear deformity is evident, while in others the ear is normal and only the jaw is affected. In more severe cases, all the structures of the first and second branchial arches can be involved. The ear, skin and underlying facial tissues such as muscles, nerves and bony structures are deficient or underdeveloped. When the eye and the spine are involved, the term Goldenhar syndrome is used. Hemifacial microsomia is the second most frequent facial anomaly, after lip and palate clefts. Studies indicated that the incidence in birth is estimated between 1 in 3500 and 1 in 5642 live births.

Case report
An eight year old boy was presented at the Department of Cranio and Maxillofacial Surgery of the University Hospital Maastricht. The child showed a severe underdevelopment of the left side of his face. The left zygomatic arch, the left ear and the left hemimandible were deficient. A broad maxillary cleft with dental agenesis was observed. The child was known to have a

Figure   5.4 Osteotomy planning on an RP model

scoliosis and finally a tetralogy of Fallot. Because of the variety of symptoms, the anomaly could be diagnosed as Goldenhar syndrome.

Distraction osteogenisis
The complexity of this case made it necessary to study the surgical options for the necessary corrections. A stereolithographic medical model of the affected skull was made. The objective of the first intervention was to correct the position of the left orbito-maxillary-complex (OMC) by distraction osteogenesis and thus narrow the cleft. Using an RP model, precise planning of the osteotomy at a Lefort III level of the left OMC was possible (Figure 5.4). The vector of movement of this bony complex was established precisely by a rehearsal operation on the RP model. A Riediger mid-face distractor, a device that allows distraction osteogenesis over a maximum distance of 20 mm, would be used. This distractor had been specially developed for indications of mid-facial advancements at a Lefort III level. The intervention was carried out according to the planning. The distraction vector was guided intraorally by an orthodontic appliance. The distance of distraction turned out to be 12.5 mm. The postoperative result showed a more favourable position of the OMC. A dramatic narrowing of the cleft was observed.

Mandibular reconstruction
The next phase of the treatment consisted of hemimandibular reconstruction. A new RP model was built for study of the design for the reconstruction of this mandibular deficiency (Figure 5.5). The model could also be used for a wax-up of the missing left mandible. The final design

Figure   5.5 Reconstruction of the mandible

showed a titanium tray for carrying slurry of tricalcium phosphate (TCP) granulates (Curasan) and bone chips mixed with PRP. The intervention was combined with removal of the Riediger mid-face distractor. A short intervention was made to fix the custom-made titanium hemi-mandible reconstruction with titanium screws. The tray was filled up with the TCP/bone mixture as planned.

Discussion
Complex cases such as patients with Goldenhar syndrome are much better understood using an RP model. Both the preoperative planning and the vector planning for the distraction osteogenesis have been made easier. The mandibular reconstruction would never have been possible without the RP model. In both interventions there was a considerable reduction in operation time.

Dr Jan Karel Th. Haex, Dr Jules M. N. Poukens and Prof. D. Riediger –
University Hospital Maastricht, The Netherlands.
                    Published in *Phidias Newsletter*, Volume 7, December 2001.

STL+ module
The Mimics STL+ module interfaces from Mimics to any kind of rapid prototyping (RP) system via triangulated files. These files are created with a bilinear and interplane interpolation algorithm to enhance the resolution of the RP model.

Standard 3D file formats such as STL or VRML (as input to virtual reality) are available. The STL format can be used by any rapid prototyping system. Powerful adaptive filtering offers a significant reduction in file size. It is possible to export from a mask, a 3D object or a 3dd file. The available export formats are ASCII STL, binary STL, DXF, VRML 2.0 and point cloud.

Several calculation parameters can be specified. STL+ makes it possible to reduce the number of triangles of the exported files, to interpolate the images and to smoothing the 3D files.

There are two methods available for reducing the number of triangles: matrix reduction and triangle reduction. Matrix reduction allows the grouping of voxels to calculate the triangles. Triangle reduction makes it possible to reduce the number of triangles in the mesh. This makes it easier to manipulate the file.

There are also two methods available for interpolating the images and generating the 3D mesh: grey value interpolation and contour interpolation. Contour interpolation is a 2D interpolation in the plane of the images that is smoothly expanded in the third dimension. Grey value interpolation is a real 3D interpolation.

A smoothing algorithm can be applied to make rough surfaces smoother.

MedCAD module
The MedCAD module provides a direct interface to CAD systems via surfaces, curves and objects exported as IGES files.

Based on the segmentation mask, MedCAD automatically generates the contours (polylines) of the mask. These polylines are used to fit b-spline curves, b-spline surfaces and objects (circle, sphere, cylinder, plane, etc.). The objects can also be created interactively. All these entities can be exported as Iges files, and are directly usable for the design of custom-made prostheses in any CAD system.

To verify the CAD implant design, Mimics imports the design as an STL file. MedCAD enables the user to visualize and manipulate the implants within the medical images in 2D cross-sections as well as in 3D.

FEA module

The Mimics FEA module makes it possible to link from scanned images to finite element analysis and computational fluid dynamics by exporting the files in the appropriate file format. 3D objects can be calculated on the basis of the scanned images and these surface meshes can be prepared for finite element analysis purposes. The remesher in the FEA module ensures that the most optimal input for that FEA package is eventually obtained. Materials can be assigned to volumetric meshes, based on the Hounsfield units in the scanned images.

The Mimics remesher significantly improves the quality and speed of FE analyses on STL models. It allows the easy transformation of irregularly shaped triangles into more or less equilateral triangles and increases the reliability and accuracy of FEA results on STL models. Most FE packages do not allow the manipulation or optimization of the mesh generated when a part is imported. This might reduce the accuracy of the results. With the remesher it is possible to optimize the file and deliver good meshes that will run in the FEA software (Figures 5.6 and 5.7).

After loading a volume mesh, the FEA calculates an appropriate Hounsfield value for each element of the mesh on the basis of the scanned images. Several Hounsfield unit ranges, each representing a material, can then be specified. A density, an elastic modulus and a Poisson's

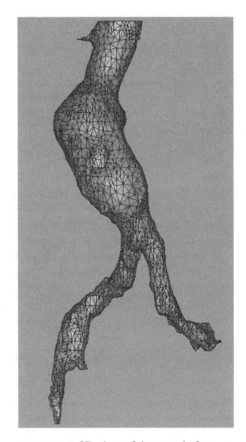

Figure   5.6 3D view of the vessel before remeshing

Figure   5.7 3D view of the vessel after remeshing

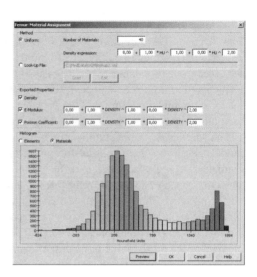

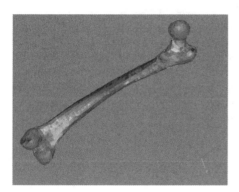

Figure 5.8 Material distribution representation

Figure 5.9 Material distribution histogram

ratio can also be assigned for those materials. The volumetric mesh with the assigned materials can then be exported to a Patran neutral, Ansys or Abaqus file (Figures 5.8 and 5.9).

Simulation module

The Mimics simulation module is an open platform for surgical simulations that makes it possible to perform a detailed analysis of data using the anthropomorphic analysis templates, plan osteotomies and distraction surgeries, or simulate and explain a surgical procedure for implant design.

To perform an anthropomorphic analysis – standard template or user defined – indicate the appropriate points (landmarks). Planes and measurements are automatically created once the points they depend upon are created. New landmarks can be created, copied, edited or deleted. Each landmark can have some default properties that can be set when creating the landmark or by editing an existing landmark. Both distances and angles can be measured. For distance, either the distance between two points or the distance between a point and a plane can be measured. As for an angle, this can be measured using three points or using two lines.

The Mimics simulation module offers a powerful 3D package for all kinds of surgery simulation application. Various tools and STL operations for simulating osteotomy and distraction surgeries are available: cut and split operations, merging of objects, mirroring of parts, distractor placement and repositioning of objects with or without a distractor. The following examples demonstrate the use of the simulation module.

Example 1: cranial implant

Cut, mirror and repositioning operations make it possible to define the required shape of a cranial implant by mirroring one side of the patient's skull (Figures 5.10 to 5.12).

Example 2: osteotomy planning

Simulation of a complex maxillary osteotomy followed by repositioning of the newly formed parts (Figures 5.13 to 5.18).

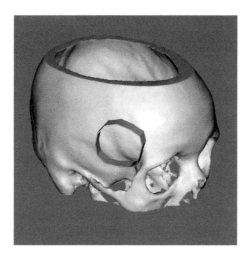

Figure 5.10 Cutting path on the intact side

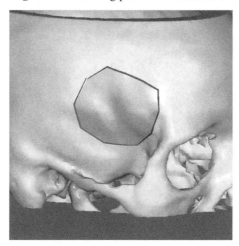

Figure 5.11 Cut and split operation

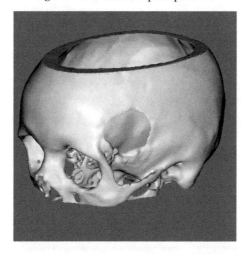

Figure 5.12 Hole with mirrored part from the intact side

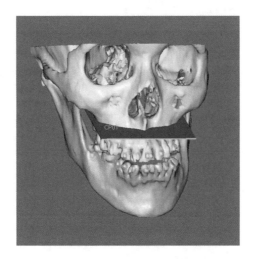

Figure 5.13 Cutting path

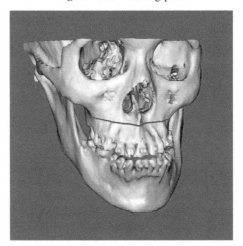

Figure 5.14 Cut operation

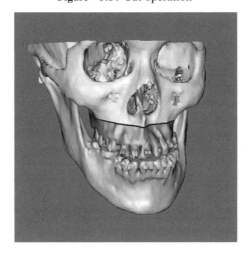

Figure 5.15 Split operation

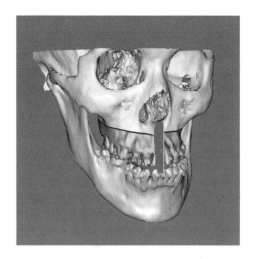

Figure   5.16 Second cutting path

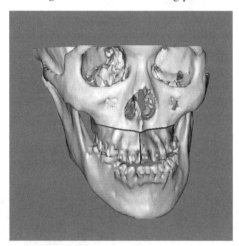

Figure   5.17 Second split operation

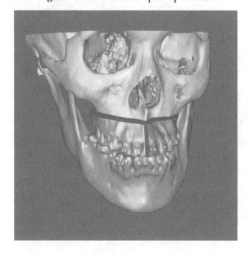

Figure   5.18 Repositioning

5.3   Computer Approach in Dental Implantology

5.3.1   Introduction

Materialise is market leader in oral implantology simulation software. Implant placement and surgery simulation packages offer increased opportunities for developing state-of-the-art techniques for minimally invasive procedures.

SimPlant® software, a software for 3D modelling and interactive 3D-based implant planning, makes it possible to create a complete planning environment from CT scanning images by reformatting of the dataset (cross-sections, panoramic image) and 3D visualization. The following crucial step in implant placement is the transfer from the preoperative plan to the surgical field. Accurate transfer of treatment planning is necessary to obtain the desired result. SurgiGuide® personalized drilling guides are designed to fit perfectly on the irregular jawbone. The irregular shape of the jawbone ensures a stable and unique positioning of the drilling guide, necessary to provide an accurate transfer of the planning.

5.3.2   Virtual 3D planning environment: SimPlant®

Although panoramic X-rays have traditionally been used to launch an implant treatment plan, many implant clinicians now realize that X-ray films are often insufficient and can even provide misleading information. A CT scan provides non-distorted data describing the patient's bone quality and quantity and hence comprehensive and accurate presurgical information.

A SimPlant® CT study creates a complete planning environment (Figure 5.19), allowing simultaneous viewing of cross-sectional, axial, panoramic and 3D images in an interactive

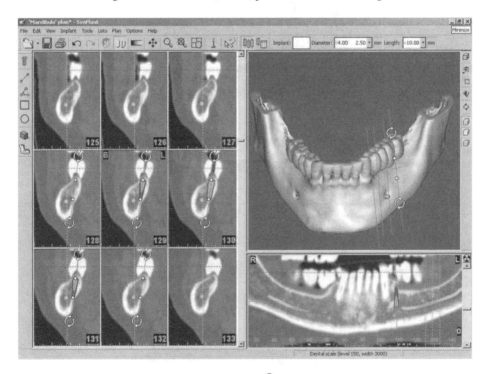

Figure   5.19 Overview SimPlant® planning environment

setting. From these, plan treatment scenarios to ensure optimal implant placement can be confidently planned.

A variety of measurement and simulation tools are included to assist in establishing the plan. SimPlant® gives access to information about bone density, measurements, sinus graft volumes, bite forces and much more. Various tools are available to visualize vital structures like the alveolar nerve channel. Virtual implants can be inserted and fully evaluated. With SimPlant®, the precise number of implants to be placed and their position, sizes and abutments will be known presurgically.

5.3.3   Guide to accurate implant treatment: SurgiGuide®

5.3.3.1   General concept of SurgiGuide®

The SimPlant® treatment plan is precisely conveyed to the SurgiGuide® and ultimately to the patient's mouth at the time of surgery.

During the operation, the SurgiGuide® is placed on the jawbone, on the mucosa or on the teeth. Owing to its precise design and intricate shape, the fit of the SurgiGuide® will be unique and stabile (Figure 5.20).

The SurgiGuide® guides the surgeon's drills to the planned position. With millimetric precision, the implant surgeon can prepare the osteotomies and place the implants with full certainty of the implant location.

SurgiGuides® are custom manufactured for each patient. A SurgiGuide® set contains consecutive guides, each with guiding cylinders following the drilling sequence specified. The consecutive SurgiGuides® fit exactly to the same position.

The production of SurgiGuides® uses stereolithography. This technology incorporates the high accuracy needed to guarantee a unique fit on the jaw.

The drill guide materials have FDA approval to be used in contact with human tissue.

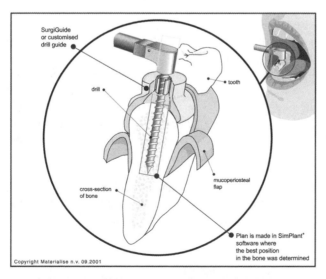

Figure   5.20 Schematic representation of a SurgiGuide®

5.3.3.2   Different types of SurgiGuide®

Tooth-supported SurgiGuide®
A tooth-supported SurgiGuide® fits on the soft tissue and teeth of the patient's partially edentulous jaw. This type of guide can be used when one or a few teeth are missing.

In combination with the implant treatment plan, a plaster cast accurately representing the patient's preoperative situation is needed to create the SurgiGuide®. This method can overcome limited distortion from scattering of artefacts which is often present owing to amalgane dental fillings or non-removable metal appliances.

Bone-supported SurgiGuide®
A bone-supported SurgiGuide® fits on the jawbone of the patient. This type of SurgiGuide® can be used for patients with edentulous and partially edentulous jaws.

Bone-supported SurgiGuides® are used when there is sufficient supporting bone surface to guarantee a stable and unique position of the guide. The fabrication of this type of SurgiGuide® requires that there is not too much residual metal (bridges, metal crowns, etc.) causing scatter artefacts on the scanner images.

During surgery, a ridge incision is made and mucoperiosteal flaps are raised to free the bone surface. The SurgiGuide® is placed on the bone surface and guides the drill in the planned position. The raising of the mucoperiosteal flaps also enables a good visibility during operation.

Case study of a bone-supported SurgiGuide®: the use of implant simulation in prosthetic rehabilitation of a class II maxilla
Dr Thierry Claeys – maxillofacial surgeon, Belgium – and Lth. Johan Kint [1] – prosthodontist, Belgium – work closely together as a team, combining their skills to produce the best possible surgical, aesthetical and functional results for their patient.

Their philosophy is that implant treatment should begin with the patient's prosthetic needs. After the type of prosthetic reconstruction is chosen, the appropriate number of implants is determined. Then, with the help of different imaging modalities, Dr Claeys and Lth. Kint evaluate the patient's jaw to determine the available bone quality and volume.

The following case illustrates their methodology.

The patient was a 67 year old doctor whose maxilla had been totally edentulous for more than 20 years. It was the desire of the patient to have a stable and retentive prosthetic reconstruction without prior surgical intervention.

To obtain the final treatment plan, the following examinations were performed:

1. Temporary teeth set-up. The wax set-up in the articulator showed a large intermaxillary gap in the distal region. Based on this and the fact that the anatomy of the edentulous maxilla was flat with a shallow palate, they concluded that it was not possible to create a fixed prosthesis in the distal area.

2. Medical imaging. Although the clinical examination and the panoramic radiographs showed adequate bone volume, the CT scan revealed an atrophic maxilla. Horizontal and vertical bone resorption was evident in the anterior, and the alveolar height in the posterior was also inadequate for implants.

---

[1] Lth. = licentiaat tandheelkunde (Dutch *dentist*).

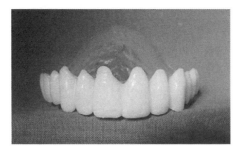

Figure 5.21 Scan prosthesis

3. Implant simulation. Since the final location and angulation of the implants determine the aesthetical and functional results, it was decided that a SurgiGuide® drill guide would be required to achieve precise implant placement in a predictable fashion. Based on the temporary teeth set-up, a radiopaque CT scan appliance was made (Figure 5.21). The patient was then scanned on a CT scanner with the appliance in place. Then the CT data were processed to produce a SimPlant® study that included 2D and 3D images of the maxilla with the desired location of the teeth to be restored provided by the CT scan appliance. Dr. Claeys and Lth. Kint then used SimPlant® software to simulate possible treatment planning solutions, choosing the most ideal treatment plan for their patient (Figure 5.22).

The SimPlant® treatment plan was then used to produce a SurgiGuide® (Figure 5.23) drill guide that would fit perfectly on the patient's maxillary bone, with drill guides accurately to position each implant exactly as it had been simulated in the SimPlant® treatment plan.

The implants were quickly and easily placed using the SurgiGuide® and the healing abutments were immediately positioned (Figure 5.24).

After an integration period of 6 months, the implants were uncovered and the prosthesis was installed. Because of the accurate placement of the implant, the prosthesis could be created with the normal prosthetic lab techniques.

Dr Claeys and Lth. Kint use SurgiGuides® when:

- accurate and predictable location and angulation of multiple implants are required;
- adequate bone volume is doubtful.

Figure 5.22 SimPlant® treatment plan

Figure   5.23 Jaw model with a SurgiGuide®

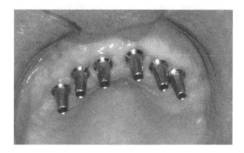

Figure   5.24 Surgery outcome

Dr Claeys and Lth. Kint concluded that, when using a SurgiGuide®, based on a computer-based treatment plan, the placing of the implants is more accurate and predictable and the surgery is simpler.

> Dr Th. Claeys, maxillofacial surgeon, St Elisabeth Ziekenhuis, Zottegem, and Dr
> J. Kint, prosthodontist – dental treatment under general anaesthesia, St Elisabeth
> Ziekenhuis, Zottegem
> Published in *DentistNews*, Volume 6, September 2002 (Apollonia cv, Ieper, Belgium).

Mucosa-supported SurgiGuide®

A mucosa-supported SurgiGuide® is a custom-manufactured drill guide made for a unique and stable fit on the soft tissue of the patient's jaw. This type of SurgiGuide® can be used for patients with edentulous jaws. A scan of the patient together with a scan prosthesis is obligatory. This also visualizes clearly the desired tooth set-up in the CT images for improved implant planning.

During surgery, the SurgiGuide®, placed on the soft tissue in the unique and stable position for which it was created, guides the drill in the planned position. This technique leads to minimal invasive surgery.

Case study of a mucosa-supported SurgiGuide®: immediate loading of mandibular
implants in a case with bleeding tendency using the mucosal-supported SurgiGuides®
The current trend in implant dentistry is to provide the patients with immediate function with dental implants. In order to do this, we have to prepare the patients for precision surgery.

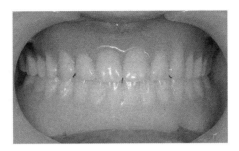

Figure 5.25 Scan prosthesis

With the development of computer-guided implantology, the preparation work for immediate loading of dental implants is simplified.

The following case report illustrates the extra benefit for a patient with bleeding tendency to receive immediate loading of the mandibular implants using the mucosal-supported SurgiGuides®.

The patient, a 62 year old lady, was known to have cardiac arrhythmia. She had undergone pacemaker implantation and was currently on anticoagulant therapy. She was wearing complete upper and lower dentures for her edentulism. However, she was suffering from pain and inadequate masticatory function owing to the instability of her lower denture.

Considering her medical condition, we decided to perform flapless surgery for implant placement in her mandible and to provide her with a fixed provisional prosthesis for immediate loading (the Hong Kong bridge protocol).

For the preparation, a duplicate of her lower denture had been made with $BaSO_4$ and it was used as the radiographic stent during CT scan investigation (Figure 5.25). The CT scan data were converted to computer images and then virtual model surgery was carried out accordingly (Figure 5.26). A series of mucosal-supported SurgiGuides® was produced, and subsequently a provisional prosthesis (Hong Kong bridge) was fabricated with respect to the model surgery (Figure 5.27).

On 12 September, 2002, the patient underwent the implant surgery under intravenous sedation with Midazolam. Two Replace Select Tapered® and two Replace Select Straight® implants with TiUnite® surface were placed in the interforaminal region of the anterior mandible without raising a muco-periosteal flap (Figure 5.28). Immediately, the final abutments were connected and the provisional prosthesis was converted to a screw-retained fixed prosthesis by bonding to the temporary cylinders with cold-cured acrylics (Figures 5.29 and 5.30).

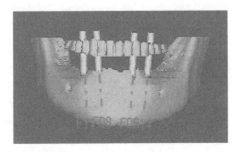

Figure 5.26 SimPlant® treatment plan

Figure 5.27 SurgiGuide®

Figure 5.28 Surgery with SurgiGuide®

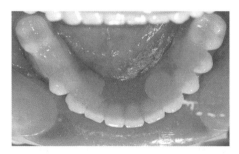

Figure 5.29 Provisional prosthesis

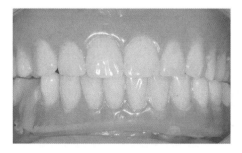

Figure 5.30 Provisional prosthesis

Postoperative recovery was uneventful and the patient was ready for the making of the final prosthesis.

Dr J. Chow Kwok Fai, oral and maxillofacial surgeon (private practice), Honorary Assistant Professor in Oral and Maxillofacial Surgery, The University of Hong Kong, pioneer in Asia for immediate loading for dental implants.

SurgiGuide® for special implants

This type of SurgiGuide® (Figure 5.31) is a custom-manufactured drill guide used for placement of special implants, like zygoma implants. Both bone-supported and mucosa-supported SurgiGuides® for special implants can be manufactured. A good positioning and small angle deviation of these types of implant is extremely important. The preoperative planning is translated to the surgery by the use of a SurgiGuide®, obtaining the best possible results for implant placement.

A model with coloured planned implants will be delivered together with the SurgiGuide®. In this way, the surgeon can check and verify the planning and the fit of the SurgiGuide®.

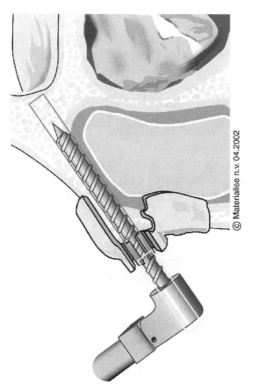

Figure   5.31 SurgiGuide® for special implants

5.3.3.3 Immediate Smile™: temporary prosthesis for truly 'immediate' loading

Typically, patients wait several months before a fixed prosthesis is placed. During this time, they will be faced with either the same deficiencies as before or with the impracticalities of wearing a conventional removable prosthesis. The onset of one-stage surgery and immediate loading, however, may have an important impact on the attitude towards implant placement by drastically reducing the time necessary to fit the patient with a fully functional and aesthetic restoration.

Several authors have already demonstrated the feasibility of successful immediate loading via an implant-retained fixed prosthesis. Moreover, there is evidence that the success rate of immediate loaded implants is similar to that obtained in cases with delayed loading, after osseointegration has taken place.

Clinicians, making use of common and well-known techniques and exploiting fully the advantages offered by CT-based implant planning and SurgiGuide®-moderated implant placement, can go about manufacturing a temporary, implant-supported prosthesis prior to the surgical intervention, thus enabling truly 'immediate' loading.

Case study of immediate loading of implants: flapless implant insertion and immediate occlusal loading of TiUnite® implants in the grafted upper jaw using a SurgiGuide®
Case report
The patient was a healthy 62 year old man who had become edentulous in the upper jaw owing to periodontal deterioration, and with only six teeth left in the anterior region of the mandible. Bilateral in the upper jaw there appeared to be no bone for implant placement under the maxillary sinuses. Therefore, the patient underwent a bilateral sinus floor elevation procedure with autologous bone taken with a trephine drill out of the anterior iliac crest.

Six months later, a CT scan (multislice multidetector high-speed spiral scan, GE) was taken with a complete radiopaque scan prosthesis in the mouth. This scan prosthesis clearly outlines the gingival. The ideal implant position is indicated by hollow sleeves in the prosthesis. A stable positioning of the scan prosthesis is a prerequisite. An occlusal splint can help to stabilize the scan prosthesis during the CT taking.

The CT data were prepared for use by the SimPlant® software (Materialise) to plan the exact position of the implants. The position, the diameter and the length of the implants were determined in the SimPlant® software, taking into account bone quality and quantity, and aesthetic considerations (Figure 5.32).

Following these guidelines, a mucosa-supported SurgiGuide® is produced.

The gingiva is punched out locally at the implant sites. The mucosa-supported SurgiGuide® is first gently stabilized with three osteosynthesis screws, avoiding compression of the mucosa. Drilling is performed according to position, angle and depth of the implants. The customized drills used have a physical stop to provide depth control. The length depends on implant length and the thickness of the gingival (Figure 5.33).

A flapless implant site preparation was performed. In total, eight TiUnite Branemark implants with a diameter of 3.75 mm were placed. The two most distal implants had a length of 15 mm, and the six anterior implants a length of 13 mm. All implants were stable and the torque measurements ranged from 10 to 35 N cm, with only two below 30 N cm (Figure 5.34).

The patient received a cross-arch screw-retained provisional bridge made of acrylic teeth on a CoCr metal frame. To allow passive fit between bridge and implants, non-engaging temporary

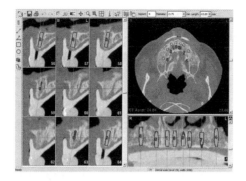

Figure 5.32 Overview implant placement

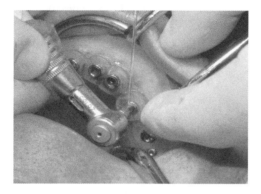

Figure 5.33 Drilling with SurgiGuide®

Figure 5.34 Implants in place after surgery

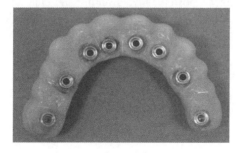

Figure 5.35 Non-connected bridge

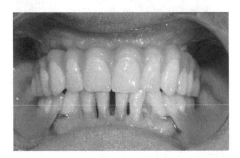

Figure   5.36 Bridge in occlusion

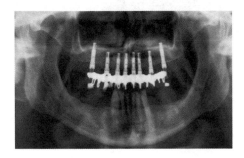

Figure   5.37 OPG with implants

titanium cylinders were mounted on the implants at implant level and subsequently connected to the bridge. The bridge was prepared in the dental laboratory on the basis of the teeth set-up used earlier as a reference to make the scan prosthesis (Unident, Leuven) and had holes where the titanium cylinders were to be connected. These holes had a certain amount of tolerance to ease seating of the bridge over the cylinders and to facilitate the occlusion (Figure 5.35). The cylinders were then connected to the bridge in occlusion with the antagonistic jaw, using autopolymerizing resin, and finished to allow interproximal cleaning of the implants.

Finally, the bridge was tightened to the implants and the occlusion checked (Figure 5.36).

Conclusion
Owing to the flapless surgery, postoperative swelling and pain are limited. The computer-aided planning and subsequent SurgiGuide® template make it possible within hours to place an aesthetic fixed prosthetic restorative reconstruction (Figure 5.37).

Dr Johan Abeloos, MD, DDS, DMD, FEBOMS, and Dr Lieven Barbier, DDS, PhD.

5.4   Conclusions

Since no two individuals are perfectly alike, a personalized approach in the medical field is a predictable evolution in treatment planning and surgery.

The above example, with the application of computer-simulated dental implant planning in Simplant®, linked to customized guides (SurgiGuides®) for a smooth and watertight transfer to surgery, illustrates that this is a perfectly possible approach. In the application field of oral implantology, this approach is now state of the art. Constant improvements to the software and types of guide ensure that its use is beneficial even in the most recent and advanced developments in implantology. Moreover current developments in modern implantology are being made possible because of the application of dental planning software and accurate guiding systems.

With this application field as a model, a field where anatomical, functional and aesthetic considerations are extremely important, we can predict similar evolutions in other medical application fields.

Bibliography

Benjamin, L. S. (2002) The evolution of multiplanar diagnostic imaging: predictable transfer of preoperative analysis to the surgical site. *J. Oral Implantology*, **28** (3), 135–44.

Ganz, S. D. (2001) CT scan technology: an evolving tool for avoiding complications and achieving predictable implant placement and restoration. *Int. J. Oral Implantology*, **1**, 6–13.

Gateno, J., Teichgraeber J. F. and Xia, J. J. (2003) Three-dimensional surgical planning for maxillary and midface distraction osteogenesis. *J. Craniofac. Surg.*, **14** (6), 833–9.

Gateno, J., Xia, J., Teichgraeber, J. F., Rosen, A., Hultgren, B. and Vadnais, T. (2003) The precision of computer-generated surgical splints. *J. Oral Maxillofac. Surg.*, **61** (7), 814–17.

Jacobs, R. and van Steenberghe, D. (1998) Radiographic Planning and Assessment of Endosseous Oral Implants, Springer-Verlag, Berlin.

Kobayashi, M., Fujino, T., Nakajuima, H. and Chiyokura, H. (1993) Significance of solid modeling of the skull using laser-curable resin in simulation surgery. *Eur. J. Plast. Surg.*, **16**, 47–50.

Kraut, R. A. (1998) Interactive CT diagnostics, planning and preparation for dental implants. *Implant Dentistry*, **7** (1), 19–25.

Lo, L. J., Marsh, J. L., Vannier, M. W. and Patel, V. V., (1994) Craniofacial computer-assisted surgical planning and simulation. *Clin. Plast. Surg.*, **21**, 501–16.

Rosenfeld, A. L. and Mecall, R. A. (1996) The use of interactive computed tomography to predict the esthetic and functional demands of implant-supported prostheses. *Compendium*, **17** (12), 1125–46.

Sarment, D. P., Al-Shammari, K. and Kazor, C. E. (2003) Stereolithographic surgical templates for placement of dental implants in complex cases. *Int. J. Periodontics and Restorative Dentistry*, **23** (3), 287–95.

Swaelens, B. and Kruth, J. P. (1993) Medical applications of rapid prototyping techniques. Proceedings of the Fourth International Conference on Rapid Prototyping, 107–20.

Tardieu, P. B, and Vrielinck, L. (2003) Computer-assisted implant placement. A case report: treatment of the mandible. *Int. J. Oral and Maxillofac. Implants*, **18** (4), 599–04.

Tardieu, P. B. and Vrielinck L. (2002) Implantologie assistée par ordinateur, cas clinique: mise en charge immédiate d'un bridge maxillaire avec des implants à appuis zygomatiques et ptérygoïdiens. *Implantodontie*, **46**, 41–8.

van Steenberghe, D., Malevez, C., Van Cleynenbreugel, J., Serhal C. B., Dhoore, E., Schutyser, F., Suetens, P. and Jacob, R. (2003) Accuracy of drilling guides for transfer from three-dimensional CT-based planning to placement of zygoma implants in human cadavers. *Clin. Oral Implants Res.*, **14** (1), 131–6.

Vander Sloten, J. (2000) *Computer Technology in Biomaterials Science and Engineering*, John Wiley & Sons, Ltd.

Vander Sloten, J., Degryse, K., Gobin R., Van der Perre, G. and Mommaerts, M. Y. (1996) Interactive simulation of cranial surgery in a computer aided design environment. *J. Cranio-Maxillofacial. Surg.*, **24**, 122–9.

Vannier, M. W., and Marsh, J. L. (1996) Three dimensional imaging, surgical planning, and image-guided therapy. *Radiol. Clin. North Am.*, **34** (3), 545–63.

Vrielinck, L., Politis, C., Schepers, S., Pauwels, M. and Naert, I. (2002) Image-based planning and clinical validation of zygoma and pterygoid implant placement in patients with severe bone atrophy using customized drill guides. Preliminary results from a prospective clinical follow-up study. *Int. J. Oral and Maxillofac. Surg.*, **31**, 7–14.

6

BioBuild Software

Robert Thompson, Dr Gian Lorenzetto and Dr Paul D'Urso

6.1 Introduction

The science of anatomical replication, referred to as biomodelling, or medical modelling, is enabled via the union of 3D medical imaging with rapid prototyping (RP), or 3D printing (3DP). Specifically, biomodelling is the science of converting scanned morphological data into exact solid replicas via specialized software and digital layer based freeform fabrication systems. BioBuild from Anatomics is such a software package designed specifically for the particular data processing requirements of biomodelling. What has always differentiated biomodelling from most RP in terms of data processing is that it must utilize a stack of 2D medical scan sections as its input data source, as opposed to the traditional CAD data source, which is usually inherently 3D in representation. This creates specific requirements relating to data acquistion, import and processing, and ultimately the quality of resultant physical biomodels. These issues and how to manage them were identified and addressed during clinical research into biomodelling throughout the 1990s, and led to the development of BioBuild by Anatomics. A brief historical overview of foundation technology may give the reader a perspective on the key developments of biomodelling. An examination of the existing BioBuild software and its utility for common biomodelling tasks, followed by future enhancements being developed for the software, will be the focus of this section.

The continuous incremental increase in computer processing power available for both patient data acquisition and subsequent image processing and biomodel production has helped make biomodelling increasingly more practical, and thus increasingly relevant to surgical practice in the twenty-first century. The simultaneous but independent development of high-resolution CT scanning and RP technology made current-generation biomodelling technically feasible, about 10 years after it was first conceptualized. CT scanning was introduced in 1972 [1], although applications in 3D imaging for surgery did not emerge in clinical practice until the system was sufficiently advanced by the early 1980s when researchers took advantage of new

Advanced Manufacturing Technology for Medical Applications Edited by I. Gibson
© 2006 John Wiley & Sons, Ltd.

hardware and software [2–4]. Alberti [5] first proposed the concept of producing physical models from CT scans in 1980, but the technologies available to process the anatomical data and then produce the biomodels were both very limited. Attempts by researchers in the 1980s to create biomodels predated RP technology, and thus the results were rudimentary, although often still considered useful for some surgical planning. Methods ranged from the stacking of life-size aluminium or polystyrene cut-outs of CT slice bone contours [4, 6] to the use of three-axis computer numerically controlled (CNC) milling to create two-part moulds for master prostheses and model castings, as described in the 1982 White technique [7]. The advent of five-axis CNC milling improved accuracy and allowed more complex biomodel construction without moulding [8]. The resultant models were of sufficient resolution to be useful for surgical planning in complex cases. However, it was evident that the complex geometries of anatomy were not ideally suited for even five-axis machining, particularly for replicating internal structures and thin walls. The required high-resolution 3D CT scanning also presented significant challenges, with concern regarding high radiation doses and long image acquisition and reconstruction times limiting the use of the 3D imaging necessary for biomodelling.

The applicability of volumetric medical imaging to surgery was to be greatly enhanced with the introduction of slip-ring spiral CT scanning in 1987 by Siemens and Toshiba [9, 10]. This enabled high-speed volumetric imaging with acceptable resolution for practical 3D imaging for the first time in clinical radiology. High-resolution (1.0 mm slice spacing) CT volumes could now be acquired with relative ease. Clinicians also benefited from the advent of IV-contrast enhanced CT angiography via dynamic scanning, with scans at 0.5 mm resolution allowing the visualization of fine cerebral vessels, previously only visible via traditional invasive angiography procedures. Simultaneous advances were also made in 3D magnetic resonance (MR) angiography. 3D reconstructions from both of these scanner types would soon establish themselves as routine radiology options for surgeons by the mid-1990s [11]. Radiation doses were also controlled via improvements in CT X-ray detector sensitivity. Additionally, the ability to space scans arbitrarily to achieve fine contiguous or overlapping slices retrospectively *after* a spiral scan block acquisition, not *during* acquisition as was previously the case with axial scanning, allowed for a 3D scan with a dose lower than its non-spiral equivalent. This also allowed for multiple reconstructions of a single spiral data block at differing slice spacings. Spiral CT scanning basically made 3D imaging 'radiology department friendly' to acquire, and significantly easier to process via related developments in 3D image processing workstation systems. Such imaging workstation systems became common accompaniments to spiral scanning systems, moving 3D imaging out of the research labs and into the normal clinical environment.

The emergence of 3D rendering techniques for voxel-based data in the 1980s allowed the shaded surface display of image volumes to demonstrate anatomy in a life-like 3D view, usually via surface shading with a virtual light source [3, 12]. Volume rendering techniques were also developed that allowed visualization of volumes via ray casting processes utilizing different rendering algorithms, without the need for surface extraction, as the volume as a whole is rendered [12–14]. These approaches examine every voxel in the volume, and were very computationally expensive at the time, with expensive specialized graphics hardware required for graphics processing. This usually limited volume rendering techniques to high-end graphics workstations in research environments.

In surface rendering techniques, the desired structure or intensity range is first delineated from the image volume via a simple threshold operation. 'Threshold segmentation' creates the object to be rendered from all voxels whose values are greater than or equal to a user-supplied threshold value. The user normally determines the intensity threshold by empirical inspection of the image volume, typically to isolate one tissue type from the others. Once the desired greyscale density values are identified via segmentation, the surface of the object must be described. This can be done via surface 'tiling' between adjacent slice contours [3], or by forming a polygon surface from exposed faces of individual voxels. This latter approach was first described by the 'marching cubes' algorithm [15], and was devised to perform surface rendering on an image volume, representing the object via a triangulated surface mesh. Such a triangulated mesh was virtually identical to the surface mesh used to describe CAD objects for stereolithography via the STL (stereolithography file). In their 1987 paper, Lorensen and Cline identified that their methods 'use polygon and point primitives to interface with computer-aided design equipment' [15]. As most CAD systems supported the new STL format for RP model creation, a software interface between 3D medical image volumes and SLA had been created as a byproduct of this landmark object surface description technique.

These developments in volumetric image acquisition and 3D image processing occurred in parallel with each other, alongside the release of the first commercial rapid prototyping system in 1986 [16]. The new stereolithographic apparatus (SLA) allowed submillimetre layer based fabrication of arbitrarily complex shapes, thus bypassing the toolpath and other restrictions of CNC milling.

The first report of the use of 3D CT with SLA for biomodelling from 1990 [17] identified SLA as superior to milling for biomodelling, and importantly recognized the similarity between CT slice data and the SLA build layer 'laser hatch' data necessary to generate a SLA model layer by layer. Mankovich chose automatically to segment the CT slices to isolate the desired bone contours, and transfer the contours directly to the SLA, after adding the required laser hatching information to each layer. As each SLA layer was 0.25 mm thick, each 2.0 mm CT slice had to replicated 8 times to allow the biomodel to be built up from this data. Problems were encountered relating to the lack of interpolation of the CT data owing to this slice replication, and also, in the physical part, support requirements for the 'contour stack'.

Klein published a paper in 1992 [18] that compared medical biomodels from milling and SLA. He identified SLA as superior, but with problems of cost and computer processing time. To counter this, he described the use of the marching cubes algorithm to create a triangulated surface description of the object. Further work was done on the contour-based (or so called 'direct layer interface') approach as explored by Mankovich in the early 1990s by Belgian company Materialise [10] which first utilized an algorithm to produce a 'stack' of interpolated RP contours in a 'stereolithography contour' (SLC) file, based on the original 'stack' of 2D images making up the 3D CT volume. This algorithm eliminated the need for any 3D surface description as well as solving the 'in between' Z plane slice interpolation problem by using cubic interpolation. It was implemented in the CT modeller (CTM) module of the Mimics software from Materialise in 1992. This is currently part of the 'RP slice' module. The major benefit of this approach was that large image volumes could be processed at high resolution, and the resultant contour files always described a 3D anatomical object more efficiently than did a STL surface file of equivalent resolution. Large triangle numbers in STL files, and long processing times required to create the necessary stack of layers from STL files for building via 'slicing' algorithms, also made the contour interface more attractive. However, the direct interface via contours approach of CTM meant that the object always had to be built in the same orientation as how it was scanned, as the contours once generated could not be rotated to optimize build height. Using the contour format also meant there was a requirement for special support structure generation software.

Australian researchers, led by Barker [19], described a process in 1993 that used powerful 3D medical imaging tools in the form of Analyze software [13] developed by the Biomedical Imaging Resource, Mayo Clinic (Rochester, USA). This approach entailed using the volumetric imaging tools of Analyze for image editing and processing, isolating the anatomical structures for biomodelling via threshold segmentation and object connectivity algorithms, then outputting the object for SLA in the STL surface format via the marching cubes algorithm. This allowed advanced 3D visualization via voxel gradient shading volume rendering [20], volumetric editing and object connectivity algorithms. This also gave users full access to the comprehensive suite of imaging functions available in Analyze. The technique was further validated for accuracy by Barker *et al.* [21] and then D'Urso [22] in separate accuracy studies.

The Barker technique was then utilized by the Brisbane biomodelling group [23] to process cases, leading to the development of cranio-maxillofacial applications as reported by D'Urso *et al.* [24], Arvier *et al.* [25] and Yau *et al.* [26]. As this data processing technique utilized the marching cubes algorithm to produce STL surface mesh files, the resultant RP build file sizes were very large owing to the triangle count created by the algorithm. This produced significant overheads relating to visualization and preprocessing for SLA, even on high-end UNIX graphics workstations used at the time. D'Urso, however, recognized the processing advantages in using the advanced volumetric tools of Analyze in combination with the layer-based contour interface to SLA. D'Urso thus adapted the Barker technique to use Analyze on the 'front-end' to interface to medical imaging, and thus enable 3D visualization and image processing, combined directly with CTM on the 'back-end' to interface to SLA and produce the smaller, more efficient SLC contour files [22]. Furthermore, D'Urso also developed a system for optimizing image volumes, allowing scans to be exported to the contour interface at an optimized orientation in terms of physical build height, as well as file size. Using Analyze in this fashion allowed the Brisbane biomodelling group to identify the necessary image processing toolkit required specifically for biomodelling. This would lead to the tailoring of a biomodelling imaging toolkit, and culminate in the development of stand-alone specialized software package designed specifically for biomodelling. That software would iteratively become BioBuild.

During the initial software development period, RP build files and biomodels produced with the new software were benchmarked against those produced via the established process, to ensure continuity of biomodel quality. Analyze continued to be used as the interface to imaging while the critical image resampling, volumetric rotations and RP file generation functionality was implemented. These 'back-end' functions were developed in a cross-platform environment, to support both the SGI UNIX platform (IRIX), as well as 32 bit Windows. Consequently, the core volumetric processing engine of BioBuild is able to utilize high-end multi-CPU workstations on both Windows and UNIX platforms. Then the required image processing functions were developed and tested in conjunction with a user interface designed specifically for biomodelling. Build height optimization [27] via automatic volumetric rotation was also implemented specifically to reduce biomodelling build times. Support for both surface STL and contour SLC output was included in the software, as, although the contour interface was more optimal, the STL interface had become dominant. This was due largely to its prevalence in RP and CAD in general, where it had become a de facto standard. The ability to manipulate such files more readily than contour files, the format portability across RP machine types and the continual increase in computing power available to process the larger STL files also contributed to its dominance. This increase in processing power and memory available in commodity hardware also allowed the deployment of the software on a PC platform in the 32 bit Windows environment, making it more accessible and affordable.

6.2   BioBuild Paradigm

The BioBuild software was specifically developed to integrate all the functionality required to import, visualize, edit, process and export 3D medical image volumes for the production of physical biomodels. An overriding design goal has always been compatibility with as many different medical scanners, and as many different image formats, as possible. Consequently, BioBuild is able to bridge the gap between patient scans and physical biomodels in many varying environments. This has seen BioBuild used in conjunction with CT data [28, 29], magnetic resonance imaging [30] 3D ultrasound data [31] and 3D angiography data.

The general process for producing a physical biomodel from a patient scan is straightforward. First, the patient dataset is imported. The data is then converted into a volume for inspection and processing. After completing the necessary processing, a 3D surface is extracted that represents the physical biomodel. To ensure that all regions of interest have been correctly modelled, it is important first to visualize and inspect the surface. Finally, the software model must be exported to a format suitable for physical biomodel production. Anatomics BioBuild software was designed to integrate each of these steps into a single user-friendly system. Further, these steps form the basis of the BioBuild paradigm for biomodel production.

The process is complicated by the many varying data formats and vendor-specific peculiarities that arise when loading data from varying scanners. There are also potentially many different biomodelling applications. Because of this, BioBuild provides the user with a vast array of options for modifying and transforming a volume. Generally, however, only a few of those operations are ever likely to be performed on any single volume. The number of possible options can initially intimidate novice users, as there is often more than one way to perform any given task. However, once novice users become accustomed to the BioBuild paradigm, they quickly become comfortable with the interface.

The usual procedural steps required for processing image volumes using BioBuild can be summarized as follows:

- import and reduce volume, and confirm orientation;
- inspect anatomy and find intensity threshold;
- edit and optimize volume;
- 3D visualization;
- RP build optimization;
- RP build file generation.

Because BioBuild was designed with ease of use in mind, virtually all volume processing can be accomplished via simple point-and-click operations. Although providing advanced editing features, most of the functionality of BioBuild is accessed through toolbars composed of simple and intuitive icons. Each toolbar can be placed at the user's desired location, but will always remain on top of the volume display. This significantly reduces the learning curve for novice users, and makes common editing operations completely automatic for experienced users.

The major steps in producing a biomodel are described below, beginning with importing a dataset.

6.2.1 Importing a dataset

3D datasets can be imported into BioBuild in several ways. Data can be loaded directly from a series of DICOM or generic 'raw' image files on the local computer, on a network drive or a remote DICOM server. Loading data from many different and varied data sources is one of the strengths of BioBuild.

Figure 6.1 shows the powerful open files dialogue, which supports real-time regular expression searches on filenames, the addition and removal of custom file filters and the ability to automatically search a directory for recognizable files and provides selection feedback, such as the number of files currently selected.

When opening a dataset, it is not necessary for it to contain all critical volume information, such as voxel size or slice spacing. If these properties cannot be automatically found within the dataset itself, BioBuild will prompt the user to enter the missing information, as shown in Figure 6.2. This option needs to be used with extreme caution, however, as entering incorrect values will result in inaccurate biomodels. Such cases require confirmation of the scanning parameters from the original scan source.

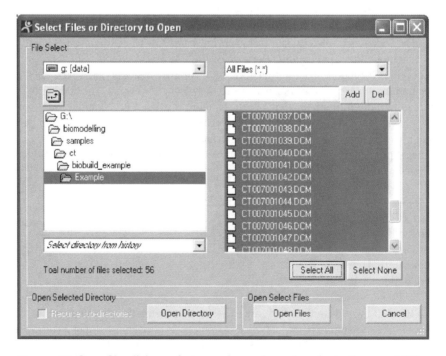

Figure 6.1 Open files dialogue features advanced search and selection capabilities

Figure   6.2 BioBuild can load raw image data, allowing the user to enter any missing volume properties

Data can also be loaded from a remote DICOM server, using the DICOM medical imaging network protocol. Remote server access is easily configured for any number of servers, as highlighted in Figure 6.3.

Many hospitals now use the DICOM standard for storage and transmission of patient data from multiple imaging modalities, and BioBuild is able to access these patient data seamlessly across the hospital network. Patient searches are made even easier by masks, such as the first letter of the patient's name, the patient's hospital ID or even the patient's doctor.

As DICOM is now the standard storage format for most patient data, BioBuild has been tested for compatibility against a wide variety of scanner types, as there are often vendor-specific peculiarities. Further, datasets of all resolutions and sizes are supported. Currently, a typical dataset consists of a voxel resolution of approximately 0.4 mm in both X and Y planes,

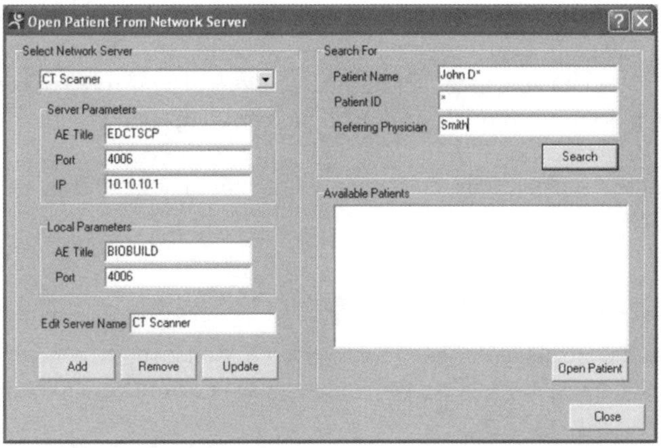

Figure   6.3 Patient data can be loaded directly from any DICOM server

and a slice spacing (Z plane resolution) of 0.3–1.5 mm. Traditionally, these were considered high-resolution values. However, with the latest model scanners, in particular multislice CT systems, such resolutions are now standard. Conversely, with higher-resolution scans comes the need for high-specification workstations to process the data in a reasonable amount of time.

The next step after loading a dataset is to inspect and edit the volume. However, before editing a volume it can be useful to reduce the overall volume size, as generally only a part of the scan will be used for biomodel production, and often scans contain a lot of unwanted information (air surrounding the head, for example). Reducing volume size can have a significant impact on processing speed, especially on low-end workstations. BioBuild has several powerful features focusing on these key aspects.

6.2.2 Volume reduction

After the desired patient study is selected, the user can preview the images via a thumbnail snapshot. The user then restricts the loaded volume size interactively by applying a clipping rectangle or freeform region that collimates the volume to include only the region of interest. This is a very useful feature for reducing overall image volume sizes and processing times, particularly when the region of interest is comparatively small to the total scan area.

6.2.3 Anatomical orientation confirmation

Once the desired scan region has been selected, and the volume reduced, the user is presented with two views of the data: the default slice data as stored in the file (usually in the axial plane) and a plane reformatted from the slice data at 90° (usually the sagittal plane). The user then scrolls through each series of images to peruse the orientation of familiar anatomical structures. If the anatomy in the images does not match the orientation annotation, the images can be flipped top-to-bottom and/or left-to-right using arrow buttons located within each window. The user then confirms the orientation by clicking a button and the series is constructed into a new BioBuild volume.

Orientation confirmation is a critical step in the biomodelling process, as scans imported into BioBuild are removed from their native environment and format. Although BioBuild is designed with this in mind, it is also imperative that a trained radiographer or imaging technician assess and confirm the scan orientation to ensure that the produced biomodel will be anatomically correct. Owing to the serious implications of a mirrored or otherwise incorrect biomodel, this step is enforced on the user and must be completed before any volume editing operations can take place.

Once confident the scans are in the correct orientation, the next critical step in biomodel production is to inspect the volume.

6.2.4 Volume inspection and intensity thresholding

The volume dataset is displayed as a 2D transverse view, in a single pane. Typically, the user scrolls through the data, slice by slice, using the mouse or arrow keys to confirm that:

- no patient movement has occurred during the scan;
- all structures of interest are included in the scan area.

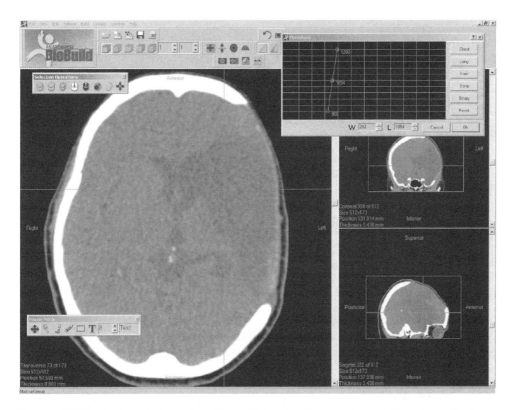

Figure 6.4 2D viewing window allows for multiple, simultaneous views of the volume. The large window on the left shows the transverse view, the upper right window shows a coronal view and the lower left window shows a sagittal view. The position of each window with respect to the others is highlighted by a green line. The separate threshold window is also visible in the top right corner, and two floating toolbars are visible

The user can repeat this process in the other two primary views (sagittal and coronal) to confirm the quality and suitability of the scan data for the intended purpose. This is important, as the quality of the physical biomodel is in large part determined by the quality of the original image volume data. The data can then also optionally be viewed in a multipaned window, displaying the three orthogonal views simultaneously – generally with transverse/axial as the main view, and coronal and sagittal secondary views. A window displaying all three views is shown in Figure 6.4.

6.2.4.1 Intensity thresholding

A crucial aspect of the inspection of greyscale image volumes for 3D imaging and biomodelling is intensity thresholding. The threshold determines the set of structures to be included in the rendering and/or biomodel. In the single or multipaned view, the intensity threshold level and window can be adjusted in real time, serving to highlight the desired structures for both 3D rendering and biomodelling. Further, contours can optionally be displayed, also updating in real time in response to threshold adjustments, which indicate regions of connectivity in the data. Contours are displayed as green lines bounding structures in the data. The contours provide an excellent aid to selecting an appropriate threshold; however, they may be disabled for a more traditional view of the image slices. The threshold is selected empirically by the user via inspection of the data and slice contours.

6.2.4.2 Display options

Options in the 2D view are available via a right-mouse click and include zooming, toggling segmentation contours on or off, toggling display smoothing on or off and opening the threshold dialogue.

Smoothing in this case refers to the blurring of the pixels in the data matrix, as illustrated in Figure 6.5. If smoothing is turned off, the individual pixels become clear. With smoothing turned on, discrete pixels are no longer discernable. The effect of smoothing is particularly noticeable when the slice data is viewed at a high zoom level.

6.2.5 Volume editing

Volume editing is the most powerful and perhaps most important feature of BioBuild, as it is the critical step in the production of a biomodel. It is this feature that allows a user to remove unwanted structures from a dataset, select the region of interest and tailor the biomodel as necessary.

Figure 6.6 shows a typical workspace set-up. All volume edits take place in the 2D slice window (left), and the final biomodel is inspected to ensure it contains all important structures (right).

The more commonly used volume editing options include the following:

- creating a 'segmentation region' via voxel connectivity at a selected threshold;
- adding and removing material;
- coloured image overlays highlighting edit areas and 'segmentation regions';
- creating a 'segmentation region' via the 'connect and keep' operation;
- selecting 'seed point' in an area of interest with the brush tool;
- selecting 'grow region' to initiate connectivity and automatically create a region containing all voxels connected to the seed point;

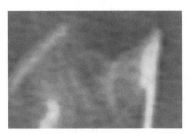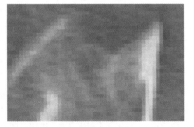

Figure   6.5 Smoothing illustration. The left image shows smoothing turned on, the right image shows smoothing turned off. The images are zoomed to 800% of normal

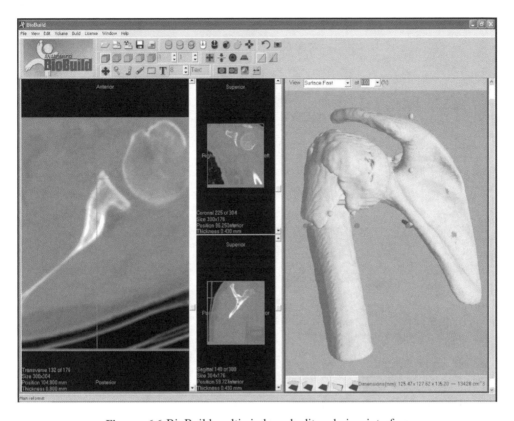

Figure 6.6 BioBuild multiwindowed edit and view interface

- deleting all voxels not connected to the current voxel selection;
- finding the smallest volume containing all selected voxels.

These, and more advanced volume editing features, are outlined below.

6.2.5.1 Connectivity options

Producing accurate and useful biomodels is closely linked to the ability to add and remove structures in the volume. BioBuild provides the powerful ability to limit or extend structures quickly and easily with its connectivity options. These include adding material to a volume to create bridges and support struts (see Figure 6.7), removing unnecessary structures and a combination of the two, which, for example, allows bone to be removed while leaving vessels intact.

6.2.5.2 Volume morphology

Volume morphology comprises the erosion and dilation of a volume, which is the expansion or contraction of a contour by a uniform mask. Consequently, the behaviour of the morphology operations is closely tied to the current threshold value. Further, *open* and *close* operations are defined. An *open* operation first performs an erosion followed by a dilation. A *close* operation

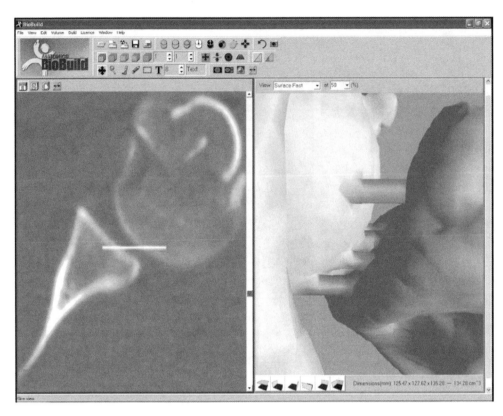

Figure 6.7 Artificial structures are easily added to biomodels. The case shown illustrates adding material to connect two bony structures. In the left window one of the artificial structures is clearly visible. In the right window the resulting structures can be easily inspected

performs a dilation followed by an erosion. Erosion and dilation have the effect of smoothing the contour lines.

Before performing morphological operations, the volume may first need to be resampled to a finer resolution. This is because BioBuild will default to the coarsest dimension to determine the mask size. If the mask is too large, the effect on the contours can be too dramatic, affecting the structure of the biomodel to an extent much greater than is usually intended.

6.2.5.3   Region morphology

Region morphology allows the user to selectively choose the structures to erode or dilate. That is, specific structures of different threshold, such as blood vessels or thin bone, can be selectively modified as desired.

6.2.5.4   Volume algebra

Often it is extremely useful to subtract a modified copy of one volume from the original. This operation, along with the union and intersection of volumes, can all be performed using the volume algebra operations of BioBuild.

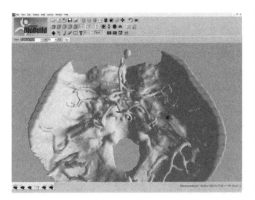 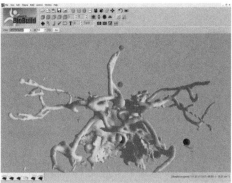

Figure   6.8 Illustration of removing bone from a model, leaving only vessels

For example, volume subtraction can be used to remove bone in CT angiography scans. This is easily achieved with the following steps. First, the volume is copied, creating two volumes of the same dataset. Then, the threshold is modified in the copied volume to select bone. A slightly low threshold is used to ensure no vessels are unintentionally removed. The bone is then selected by seeding a point on any bony structure with the paint brush tool and starting the 'region grow' operation. This causes all voxels containing bone, and connected to the seed point, to become selected. Because a low threshold was used, the bone is now dilated to compensate. The selection is then inverted and the 'remove material' operation is performed. By inverting the selection, everything *except* bone is removed. Finally, the copied and edited volume is subtracted from the original, leaving only the vessels. Figure 6.8 illustrates this process. The left image shows the original skull base including vessels. The right image shows the vessels with bone removed.

The dilation step is important as a low threshold was used initially to select the bony structures. This means that not all the bone may be included. Dilating ensures that, when the copied volume is subtracted from the original, no bone remains.

6.2.5.5   Labels

It is possible to label biomodels automatically, using the 'text' feature. By adding text to a volume, a separate STL file is created that represents the desired text. It can also be given a different intensity to differentiate it from the anatomical model. When the biomodel is built physically, the text can appear as a different colour on the biomodel, and more than suffices as a labelling technique. To allow labelling in non-colourable RP systems, the text can be punched into the model using the 'remove material' editing option, or added as a raised label using the 'add material' editing option.

6.2.5.6   Volume transformations

If a scaled or otherwise transformed model is required, BioBuild allows the user to modify the voxel dimensions and rotate the volume by an arbitrary amount in any plane. This allows

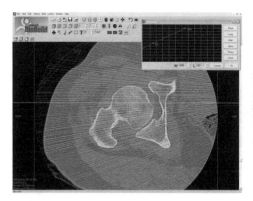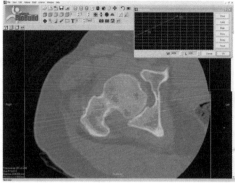

Figure   6.9 Left image shows a patient scan with extremely bad noise, severely affecting the contours. After applying the smoothing filter twice, the contours in the right image more accurately reflect the desired structures

volume inspection and editing in any plane the user requires. The volume can also be resampled at a higher or lower resolution, to suite the needs of each individual case. This results in a finer or coarser approximation of the volume. By resampling at a lower resolution, processing time can be reduced for low-end systems. Resampling at a higher resolution can help improve the final biomodel quality of a low-quality scan.

6.2.6   Image processing

To aid general editing, BioBuild also supports linear filters to smooth and sharpen images. Image smoothing can be extremely useful for the removal of noise from an image. Consider Figure 6.9. On the left is a patient scan with very bad scatter noise due to a hip replacement on the other non-visible side. On the right is the same scan, at the same threshold, after smoothing has been applied. Note that most of the scatter noise has been removed, leaving only the major structures.

This technique can be used to remove noise, as well as to remove small artefacts in an image series. Note that smoothing is applied across the entire data series, and not single images, to ensure consistency and correctness.

6.2.7   Build orientation optimization

The patented one-click build orientation optimization [27] is one of the strongest features of BioBuild, as the orientation of a biomodel has a large impact on both the time and cost of its production. This is because different orientations result in varying amounts of support materials and drastically altered build times. For example, on many RP systems, the time taken to build a biomodel can be significantly reduced if the model is built in an orientation that minimizes height. The one-click optimization allows the user to optimize the height of a build with a single click, reducing both the time to build and the support structures required.

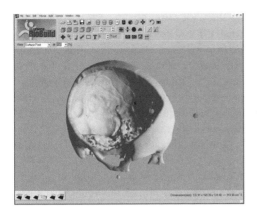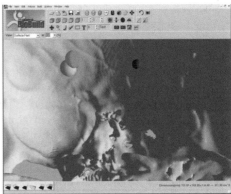

Figure   6.10 BioBuild features a fully interactive 3D visualization window, which allows the user to rotate, pan and zoom the model

6.2.8   3D visualization

After all the necessary volume editing has been completed, the user should inspect the results in a 3D surface rendering. It is vitally important that the user inspects the rendered surface at the selected threshold to ensure the suitability of the threshold, and confirm the presence of all the structures required in the physical biomodel. This is achieved using the 3D visualization capabilities of BioBuild to create a 'virtual biomodel'. In the 3D view the user can fully interact in real time with the model by panning, zooming and rotating the model into any desired position. It is possible to zoom in for extreme close-ups of the finest features in the model, including internal structures, as well as view the model from any angle.

Figure 6.10 illustrates the 3D view. The image on the left is at the default zoom level, while the image on the right shows the same model, in the same orientation, at a much higher zoom factor. The model is easily manipulated using the standard three-button mouse interface of left button for rotation, middle button for pan and right button for zoom. There are also icons for immediately moving the model into predefined orientations, such as front, back, left and right. This viewing capability makes inspecting the 'virtual biomodel' quite simple.

When generating the surface model for the 3D view, the user has the ability to select a quality setting. This in no way affects the quality of the final biomodel, but it can reduce the complexity of the viewed surface model for systems without the necessary CPU or graphics performance.

6.2.9   RP file generation

To produce a physical biomodel it is necessary to generate a file specifying the biomodel surface in a format that an RP machine can interpret. BioBuild allows the user to generate an RP file at any time, either as an STL surface or as an SLC contour file.

STL file generation defaults to creating a surface mesh at a resolution defined by the coarsest voxel dimension. This is typically the Z plane or CT scan direction. Traditionally, spiral CT scans used are 1.0–1.5 mm apart. At this resolution, some interpolation is required to improve biomodel resolution and consequently surface finish. With submillimetre slice spacings becoming more prominent with the uptake of multislice CT systems, the user can now create RP build files with an STL facet size the same as the original scan spacing, requiring no image interpolation. Scan spacings of 0.3–0.5 mm over previously unachievable anatomical distances can produce excellent results in the resultant biomodels. Of course, attention then has to be paid to large STL file sizes and their effects on system performance in these instances. This is increasingly less important, however, as computing power continues to improve.

The surface model is generated from the current volume and threshold setting. Typically, the user will iterate between volume editing and 3D visualization several times before generating a STL surface file to ensure the best outcome.

A direct contour interface to RP via the SLC format still remains more optimal in terms of file size and processing requirements, but it lacks the portability of STL and requires separate build support structure software. A user creates an SLC file from the current volume with its current threshold by simply clicking the 'build contour' button and selecting the desired layer thickness. The layer thickness selected must then correspond directly to the build layer thickness of the target RP or 3D Printing hardware. The image volume will then be interpolated to a new volume at this selected thickness, and then contours will be extracted from each interpolated slice as per the set intensity threshold. The extracted contours will then reflect the anatomical structures of interest in each layer, as defined by the threshold.

It is also possible to select multiple volumes at the same time when generating an STL or SLC file, which allows for selective colouration (as available in SLA), or batch processing. This is useful when creating RP files with different thresholds set for each volume. Labels for biomodels are constructed in this fashion.

6.3   Future Enhancements

BioBuild is continually being updated to meet the demands of clients and to make the user interface more intuitive and even simpler. Part of this redesign includes functionality enhancements such as multiple simultaneous thresholds, each with its own colour code, advanced surface generation techniques [32], direct volume rendering (DVR) techniques [33, 34] for rapid biomodel preview without the need for surface generation and new powerful 3D editing tools. Further, with the increasing size of patient scans it is imperative that all operations remain interactive and that any dataset, regardless of size, be editable in an interactive manner. To this end, the memory model of BioBuild has been optimized for arbitrarily large volumes, and the software is incorporating the latest in DVR techniques, as well as making use of the recent improvements in commodity graphics cards [33].

6.3.1   Direct volume rendering (DVR)

The goal of DVR is to render a volume in real-time without the extraction of a surface. That is, the volume data are rendered directly. This is achieved by an optical data model that maps scalar values to optical properties like colour and opacity [34], commonly termed a transfer function. Typical approaches to DVR are based on multitexture rendering, in which the volume is represented by a series of images, blended together to give the appearance of a 3D object.

Until recently, real-time volume rendering was only achievable on high-end, expensive workstations. With recent major advances in commodity graphics cards, however, it is now possible to render very large datasets on inexpensive, consumer-grade hardware. This has opened the way for DVR on the desktop. This is important for biomodelling, as the trend towards larger and larger datasets means radiographers/technicians increasingly rely on volumetric reconstructions. Thus, it is imperative that BioBuild support DVR for volumetric preview and editing.

6.4 Conclusion

BioBuild has been found to be user-friendly, stable, efficient and accurate for biomodelling applications. It supports a wide variety of processing tools, including image filters, volume morphology and algebra, patented one-click orientation optimization and export to STL and SLC formats. The interface is simple and intuitive: once accustomed to the BioBuild paradigm, most operations become automatic. All this, combined with a design centred on supporting as wide a variety of scanners and data formats as possible, makes BioBuild the perfect bridge between patient scans and physical biomodels in almost any environment.

References

Eisenberg, R. (1992) *Radiology: An Illustrated History*, Mosby Year Book, Missouri.

Hemmy, D. C., David, D.J. and Herman, G.T. (1983) Three-dimensional reconstruction of craniofacial deformity using computed tomography. *Neurosurgery*, **13**, 534–41.

Marsh, J. L. and Vannier, M. W. (1983) Surface imaging from computerized tomographic scans. *Surgery*, August, **94**(2), 159–65.

Vannier, M. W., Marsh, J. L., Gado, M. H., Totty, W. G., Gilula, L. A. and Evens, R. G. (1983) Clinical applications of three-dimensional surface reconstruction from CT scans: experience with 250 patient studies. *Electromedica*, **51**, 122–31.

Alberti, C. (1980) Three-dimensional CT and structure models. *Br. J. Radiol.* March, **53**, 261–2.

Blake, G. B., MacFarlane, M. R. and Hinton, J. W. (1990) Titanium in reconstructive surgery of the skull and face. *Br. J. Plast. Surg.*, **43**, 528–35.

White, D. N. (1982) Method of forming implantable prostheses for reconstructive surgery. US Patent 4436683, June 3.

Zonneveld, F. W. and Noorman van der Dussen, M. F. (1992) Three-dimensional imaging and model fabrication in oral and maxillofacial surgery. *Oral Maxillofac. Surg. Clin. N. Am.*, **4** (1), 19–33.

Kalender, W. A., Seissler, W, Klotz, E. and Vock, P. (1990) Spiral volumetric CT with single breath hold technique, continuous transport, and continuous scanner rotation. *Radiology*, **176**, 181–3.

SME (1997) *Rapid Prototyping Technology: a Unique Approach to the Diagnosis and Planning of Medical Procedures*, Society of Manufacturing Engineers, Dearborn, Michigan.

Aoki, S., Sasaki, Y., Machida, T., Ohkubo, T., Minami, M. and Sasaki, Y. (1992) Cerebral aneurysms: detection and delineation using 3-D-CT angiography. *Am. J. Neuroradiol.*, **13**, 1115–20.

Goldwasser, S. M., Reynolds, R. A., Talton, D. A. and Walsh, E. S. (1988) Techniques for the rapid display and manipulation of 3-D biomedical data. *Comp. Med. Imag. and Graphics*, **12** (1), 1–24.

Robb, R. A., and Hanson, D. (1991) A software system for interactive and quantitative visualization of multidimensional biomedical images. *Aust. Phys. Eng. Sci. Med.*, **14** (1), 9–30.

Drebin, R. A. (1988) Volume rendering. *Computer Graphics*, **22**, 65–74.

Lorensen, W. E. and Cline, H. E. (1987) Marching cubes: a high resolution 3D surface construction algorithm. *Computer Graphics*, **21**, 163–9.

Hull, C. W. (1986) Apparatus for production of three-dimensional objects by stereolithography. US Patent 4575330, March 11.

Mankovich, N. J., Cheeseman, A. M. and Stoker, N. G. (1990) The display of three-dimensional anatomy with stereolithographic models. *J. Digit. Imaging*, **3**, 200–03.

Klein, H. M., Scheider, W., Alzen, G., Voy, E. D. and Gunther, R. W. (1992) Pediatric craniofacial surgery: comparison of milling and stereolithography for 3D model manufacturing. *Pediatr. Radiol.*, **22**, 458–60.

Barker, T. M., Earwaker W. J. S., Frost, N. and Wakeley, G. (1993) Integration of 3-D medical imaging and rapid prototyping to create stereolithographic models. *Aust. Phys. Eng. Sci. Med.*, **16**, 79–85.

Robb, R. A. (1995) *Three-Dimensional Biomedical Imaging – Principles and Practice*, VCH Publishers, New York.

Barker, T. M., Earwaker, W. J. S. and Lisle, D. A. (1994) Accuracy of stereolithographic models of human anatomy. *Australas. Radiol.*, **38**, 106–11.

D'Urso, P. S. (1997) Stereolithographic biomodelling in Surgery. Unpublished PhD thesis, University of Queensland.

D'Urso, P. S., Barker, T. M., Earwaker, W. J. S., Arvier, J., Lanigan, M., Atkinson, R. L., Tomlinson, F. H., Loose, S., Donerly, W., Wilson, A., Weidmann, M. J., Redmond, M., Hall, B. I., Yau, Y. Y., Askin, G., Mason, S., Merry, G., Kelly, G., Frost, N., Wakeley, G. D., Riek, A. T., M'Kirdy, B. H., Stewart, J. and Effeney, D. J. (1994) The Australian SLA modelling project. Proceedings of the 2nd International Workshop on Rapid Prototyping in Medicine, April 7–8, 1994, Zurich, Switzerland, p. 8.

D'Urso, P. S., Barker, T. M., Effeney, D. J. *et al.* (1993) Stereolithographic (SLA) biomodelling in surgery, preliminary results at 18 months. Proceedings of the Surgical Research Society of Australasia, Annual Scientific Meeting, July 21–23, 1993, Brisbane, Australia, p. 24.

Arvier, J. F., Barker, T. M., Yau, Y. Y., D'Urso, P. S., Atkinson, R. L. and McDermant, G. R. (1994) Maxillofacial biomodelling. *Br. J. Oral Maxillofac. Surg.*, October, **32** (5), 276–83.

Yau, Y. Y., Arvier, J. F. and Barker, T. M. (1995) Technical note: maxillofacial biomodelling – preliminary result. *Br. J. Radiol.*, May, **68** (809), 519–23.

D'Urso, P. S. (1993) Stereolithographic modelling process. US Patent 5 741 215, September.

D'Urso, P. S., Atkinson, R. L., Lanigan, M. W., Earwaker, W. J. S., Bruce, I. J., Holmes, A., Barker, T. M., Effeney, D. J. and Thompson, R. G. (1998) Stereolithographic (SL) biomodelling in craniofacial surgery. *Br. J. Plast. Surg.*, **51** (7), 522–30.

D'Urso, P. S., Askin, G., Earwaker, W. J. S., Merry, G., Thompson, R. G., Barker, T. M. and Effeney, D. J. (1999) Spinal biomodelling. *Spine*, **24** (12), 1247–51.

D'Urso, P. S., Thompson, R. G., Atkinson, R. L., Weidmann, M. J., Redmond, M. J., Hall, B. I., Jeavons, S. J., Benson, M. D. and Earwaker, W. J. S. (1999) Cerebrovascular biomodelling. *Surg. Neurology*, **52** (5), 490–500.

D'Urso, P. S. and Thompson, R. G. (1998) Fetal biomodelling. *Aust. and New Zealand J. Obstetrics and Gynaecol.*, **38**, 205–7.

Klein, R. and Schilling, A. (1999) Fast distance field interpolation for reconstruction of surfaces from contours. In *Eurographics '99*, Short Papers and Demos Proceedings, Milan, Italy.

Rezk-Salama, C., Engel, K., Bauer, M., Greiner, G. and Ertl, T. (2000) Interactive volume rendering on standard PC graphics hardware using multi-textures and multi-stage rasterization. Proceedings of the ACM SIGGRAPH/EUROGRAPHICS Workshop on Graphics Hardware, 2000, Interlaken, Switzerland, pp. 109–18.

Engel, K. and Ertl, T. (2002) Interactive high quality volume rendering with flexible consumer graphics hardware. In *Eurographics '02*, State of the Art Report.

7

Generalized Artificial Finger Joint Design Process Employing Reverse Engineering

I. Gibson and X. P. Wang

7.1   Introduction

It is widely accepted that rheumatoid arthritis (RA) can cause severe suffering. A promising method in clinical treatment, finger joint replacement surgery, has attracted great attention in both research and clinical practice. Here, the human finger joint structure is first introduced, and one of the common diseases – RA – is then identified. Following that, finger joint replacement surgery is briefly interpreted. Research problems are then identified on the basis of an analysis of current finger joint replacement methods.

7.1.1   Structure of a human finger joint

The typical configuration of human finger joint bones is illustrated in Figure 7.1 (Strete, 1997).

7.1.2   Rheumatoid arthritis disease

Arthritis is a disease that can cause painful inflammation of joints and even breakdown of the joint structure. Arthritis, in particular rheumatoid arthritis (RA), affects hundreds of millions of patients throughout the world, causing pain, deformity, stiffness, weakness and incapacity. In particular, smaller peripheral joints such as the proximal interphalangeal joints (PIPJ) are the

Advanced Manufacturing Technology for Medical Applications   Edited by I. Gibson
© 2006 John Wiley & Sons, Ltd.

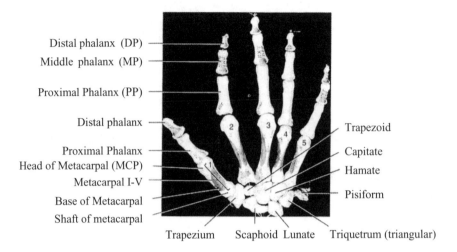

Distal phalanx (DP)
Middle phalanx (MP)

Proximal Phalanx (PP)

Distal phalanx

Proximal Phalanx
Head of Metacarpal (MCP)
Metacarpal I-V

Base of Metacarpal

Shaft of metacarpal

Trapezoid

Capitate

Hamate

Pisiform

Trapezium   Scaphoid Lunate   Triquetrum (triangular)

Figure   7.1 Human finger joint structures

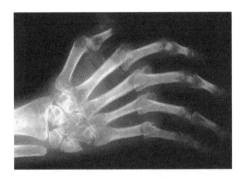

Figure   7.2 Radiograph of a hand in late-stage rheumatoid arthritis

most commonly involved joints when RA affects the hand. Figure 7.2 (Fishback, 1999) shows a radiograph of a hand in late-stage RA, illustrating the subluxation of the metacarpo-phalangeal joint (MCPJ) and ulnar drift of the fingers. It can be seen from this image that the patient suffers severe joint displacement, rendering the hand immobile.

RA affects about 750 000 people in the United Kingdom and is the biggest cause of pain and disability. In Hong Kong there are about 1.1 million people over 60 years old (from the year 2001), and 0.4% of them (4400) are likely to develop RA. About 25% (1100) RA cases are severe enough to warrant implantation of artificial finger joints. There were approximately 120 finger joint replacements carried out in Hong Kong each year from 1996 to 2000, requiring about 2–8 artificial finger joints each, depending on the philosophy of the surgeon and the extent of the damage in the patient.

7.1.3   Finger joint replacement design

During the early stages of arthritis, the disease may be treated with drugs or physiotherapeutic techniques. In later stages, synovectomy and soft tissue reconstruction may be applied to

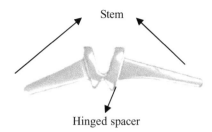

Stem

Hinged spacer

Figure 7.3 Graphics of the NEUFLEX MCP/PIP finger joint implant System

preserve joint motions and functions. However, during these later stages, there may be no alternative to performing a total joint replacement.

People who lose functions in their fingers because of RA, 'wear and tear', osteoarthritis or injury may be candidates for finger joint replacement. Finger joint replacement surgery, or finger arthroplasty, has been an optional treatment for people for more than 30 years. For example, ulnar drift (Figure 7.2) can disable patients and leave their fingers virtually useless. Ulnar drift is an abnormality of position of the MCPJ where the joints are deviated towards the ulna. This deviation is seen in patients with various disorders, including rheumatoid arthritis, systemic lupus erythematosus and Jaccouds arthropathy.

A person's finger joints act as hinges between the bones in the hands and fingers. They help our hands grip and hold things. Patients who have osteoarthritis or have experienced traumatic knuckle injuries may benefit from finger joint replacement surgery. Finger replacement surgery may help these patients regain movement and a more natural look to their hands while easing them from chronic arthritic pain.

In a finger joint replacement, doctors remove the parts of the damaged joint that cause pain and implant an artificial joint. For example, the NEUFLEX metacarpal phalangeal (MCP) and proximal interphalangeal (PIP) finger joint implants (Figure 7.3) have a stem on either end with a spacer in the middle. Doctors place the stems into the ends of the bones on either side of the joint that has been removed.

7.1.4 Requirements for new finger joint design

When the joints are destroyed, artificial joint replacement of the hip, knee, shoulder and elbow has provided excellent results in 95–99% of cases. In the case of finger joints, however, because the operation is frequently performed late when the ligaments are also involved, results have not been as satisfactory.

The initial design of various hinge prostheses led to loosening, breakage and erosion through the bones. A series of two-component prostheses did not lead to better results, either because of a problem between the bone and prosthesis or because of a much-compromised shape of the articulating surface. The third group of prostheses using the concept of flexible implant are actually spacers around which ligaments and capsules are reconstructed. The most common artificial finger joint being used in clinical situations nowadays is Swanson's silicon rubber joint spacer. However, the range of movement achieved is only 30–40° (normal: 90–100°), and most will break after a few years of use, frequently leading to fragmentation and reactive inflammation.

The search for new material faces the problem of rejection, wear and tear and subsequent particle formation in the body. For example, the use of silicon breast implants has led to so many problems that it has to be withdrawn from the market.

Furthermore, the shapes of implants pose a demand on perfection compromised by practicality. The problem of loosening in the conventional artificial joints is also still unsolved. Investigators are therefore exploring new designs with new material, new shapes and new anchorage to bones.

New designs of finger joint prostheses should have improved performance in these critical areas:

- Greater ease of motion: the finger joint prosthesis is designed to give people a greater range of motion in their fingers. Existing implants are not good at distributing the stress of fully flexing the fingers, which leads to deterioration of the implants. The finger joint implants should more closely resemble the natural structure of the hand, decreasing the stress at the hinge of the implant itself. Less stress means better range of motion, as well as less risk of fracture for the implant itself.
- More natural appearance: the finger joint implant should enable doctors to give fingers a more natural look after joint replacement surgery.
- Increased durability and functional properties: careful quality control procedures ensure the highest-quality implants.
- The joints should last longer and the patient should experience greater comfort and mobility.

7.1.5   Research objectives

The objective of the research is to design a surface replacement finger joint prosthesis with the geometry of the bearing surfaces designed as close to the original anatomy of the human finger joint as possible. This is a cooperative project with the Department of Orthopaedic Surgery in the University of Hong Kong (HKU), which aims to develop a new generation of finger joint replacement design that has the potential to be commercialized locally and internationally.

The design of the bearing surface is the most important part of the finger joint prosthesis. Previous designs only have relatively simple geometric structures. The current designs should be revised with improved surface geometry as close to the anatomical shape and function of the real human finger joint as possible to reproduce stability and a near-normal range of movement. The new design should also be subject to mechanical testing, biomechanical testing and animal testing before being used for human trial.

Computer-aided methods were used for this study using freshly frozen cadaveric finger joints. Laser scanning technology was used to create a range of shapes based on actual human bone samples for evaluation. After acquisition of the three-dimensional (3D) image data of finger joint bone using a 3D scanner, the bone surface is reconstructed. Accurate replicas of the topology of the original finger joint can then be produced. A database of the geometry and dimensions of the finger joints is constructed, and then the data are generalized to make them suitable for widespread use.

The procedure being investigated in this research for the extraction of finger joint surface geometry will be used in the design and manufacturing of the finger joint surface replacement prosthesis. The newly designed prosthesis will be far more proximal to the real bones than previous designs, with minimized impact of joint prosthesis on the body, hence making it possible to preserve hand functions as long as possible.

7.2 Supporting Literature

All finger joint prostheses that are used for finger joint replacement are designed to relieve pain, to restore a functional range of motion (ROM), to correct existing and prevent further deformity and to give cosmetic improvement. Many prostheses have attempted to achieve all these aims since the first one was implanted over 40 years ago.

7.2.1 Previous prosthetic designs

Research into the design and development of prostheses has made major advances in the last half-century, resulting in complex devices for almost all articulating joints of the human body, especially large-weight bearing joints such as the hip, femur and knee. Nevertheless, there is some frustration among hand surgeons in not being able to transfer large joint technology to the joints of the hand. This is probably because of the small sizes of the joints, their presence within kinematic chains and their complex soft tissue investments (Landsmeer, 1963; Zancolli, 1968; Valero-Cuevas *et al.*, 1999; Purves and Berme, 1980; Giurintano *et al.*, 1995; Backhouse, 1968).

Despite difficulties, many designs of finger joint prostheses have been developed for the cure of finger joint diseases, which include osteoarthritis, rheumatoid arthritis, gout, psoriatic arthritis and juvenile rheumatoid arthritis. All finger joint prostheses are designed to relieve pain, to restore a functional range of motion, to correct existing and prevent further deformity and to give cosmetic improvement. These prostheses can be grouped into one of three basic designs: linked and constrained design, non-constrained design and spacer design (Chow, 2000):

- *Linked and constrained designs.* Linked and constrained designs are manufactured entirely from metal and only allow flexion and extension. These designs invariably fail on account of the fact that forces transmitted through these joints are very high, especially torque forces when pinching with the thumb. Examples of linked and constrained designs are the Brannon and Klein, Flatt, Steffee, KY Alumina ceramic and Minami Alumina ceramic prostheses.
- *Non-constrained designs.* The uniaxial designs evolved into multiaxis non-constrained designs with snap-fit assemblies, which are manufactured from metal or ceramics and polymers. They give flexion, extension, abduction and adduction movements. However, all these have reversed the shape of the components, so that the roller or the ball part is on the proximal phalangeal side, and the cup or socket is on the metacarpal side. The range of movement has therefore never been good. Examples of non-constrained metacarpal phalangeal joint prostheses are Griffith-Nicolle, Schetrumpf and Schultz.
- *Spacer designs.* Spacer designs are made of either silicon rubber or polypropylene around which soft tissue reformations precede. These cheap, easily moulded prostheses allow flexion, extension, abduction and adduction. The range of movement has tended to be 30–60°, and ulnar deviation is possible in a large proportion of cases. Examples of spacer designs are Swanson, Niebauer, Helal and Nicolle-Calnan.

7.2.2 More recent designs

The latest generation of finger joint prostheses adopts the new design concept of surface replacement design, which will restore the anatomy of the real finger joint. The Beckenbaugh prosthesis (Beckenbaugh, 1983), the Linscheid prosthesis (Linscheid and Beckenbaugh, 1991; Linscheid *et al.*, 1997) and the Durham prosthesis (Ash and Unsworth, 2000) are examples of the new anatomical shape designs. These joint prostheses require minimal bone resection, and therefore ligaments and muscles surrounding the joint are still functional to provide joint stability. These designs allow the joint to function as naturally as possible and are especially suitable for younger patients who have high grip strength and who have a further life expectancy of 50 or more years.

Anatomical shape designs can provide pain relief, correction of deformity and improved range of motion in the short term. However, breakage, erosion, deformity, dislocation and other problems plagued many of these initiatives. The design of any medical implant requires that the host region is fully understood. For the design of finger joint prostheses, it is vitally important that an in-depth understanding of the geometry and anatomy of the contact surface is obtained. However, little focus has been placed on the complex geometry of the bearing surface of the finger joint, to provide an anatomical basis for designing the finger joint stems of the prosthesis, and the dimensions for the replacement of the contact surface.

Development in large joints is now focusing on new materials such as ceramics which avoid the problems of polyethylene particles and have minimal wear (Swanson, 1972). Beckenbaugh (1983) reported on the long-term use of a pyrolytic carbon coated finger joint prosthesis, which seemed to give good results as far as wear and tear is concerned.

Of the many different types of finger joint prosthetic replacement that have been used clinically, the most commonly used is perhaps still Swanson's prosthesis, which is one of the first finger joint prostheses designed in 1959 (Brannon and Klein, 1959) for MCPJ. However, prosthesis fractures and recurrence of deformity have still occurred, and particle synovitis has been produced owing to breakage and wear of the silicon rubber. A new prosthetic design that can obtain a near full range of movement and can overcome finger joint complications is therefore of great potential benefit.

7.2.3 Development of a new design

In the immediate future, a combination of three recent developments (anatomical shape, osteo-integration and new material) may open up a new direction in the design of finger joint prostheses. As in the development of artificial knee joints, it has finally been realized that a near full range of movement cannot be obtained unless the shapes of the components are anatomical. Osteo-integration will obviate the use of bone cement. Use of new material that is wear resistant may solve the problem of wear particles and subsequent synovitis and loosening. Chow's design (Figure 7.4) is considered to be workable but requires further consideration, particularly with respect to geometry and mechanical properties, by incorporating studies of features in actual bone samples, material choice, manufacturing methods, ease of use, etc.

To design a human body prosthesis, good results can be achieved only when accurate input information is given. Since there is little adequate information on the dimensions of the contact surfaces to create a surface replacement prosthesis to match the original joints, work was carried

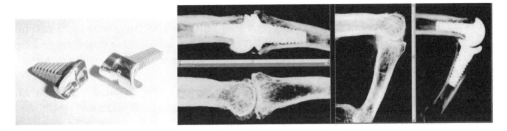

Figure   7.4 Chow's PIPJ prosthesis plus preoperative and postoperative X-ray images

out to investigate the accurate geometry of the PIPJ and MCPJ bones. These studies can be used in designing a finger joint surface replacement prosthesis to minimize the impact on the body and hence preserve hand function.

Unlike previous designs of finger joint prostheses that obtained dimensions of the finger joint surfaces using shadowgraphs or radiographs, we can now make more accurate designs based on scanned data obtained from actual finger joints. The final surface replacement could be designed closer to the original anatomy of the finger joint.

The new design of artificial finger joint is based on three key factors:

- Anatomical shape. Previous designs only have a simple geometric structure. The new design for the anatomical part will make use of laser scanning technology and rapid prototyping to create a range of shapes based on actual human bone samples for evaluation.
- Advanced materials. Investigations into ceramic fabrication methods for improvement in wear resistance in conjunction with metal for support and osteo-integration have previously not been applied in the way proposed for artificial finger joints.
- Two-piece design. The proposed design uses two pieces for each side of the joint. The stem piece provides anchorage to the bone, while the top provides articulating surfaces. This approach not only eases the difficulties concerning material selection but also improves the clinical procedure by allowing parts to be matched to suit individual patients. In cases where there is a need to change components, this design will also ease the process, requiring only the articulating surface component to be changed, leaving the bone anchorage system intact.

7.2.4   Need for a generalized finger joint prosthesis

An ideal implant must reflect a compromise between engineering and biomechanical consider-ations, material characteristics, anatomical and physiological considerations and the patient's needs. It must be pain free, mobile, stable and durable (Swanson, 1969; Swanson, 1972; Mannerfelt and Andersson, 1975).

It is hoped that, by designing the bearing surfaces of the finger joint prosthesis as close to the original anatomy of the finger joint bone as possible, the prosthesis will not change the mechanical advantage of the finger joints and will therefore put normal stresses on the soft tissues surrounding the joint. Since NURBS curves are used in the reconstruction of the finger joint for the analysis of the freeform surfaces in this study, it would be better if the curves used were as close to the original finger joint surface curves as possible. Different numbers of NURBS curves and of control points for each NURBS curve were compared for the construction of the final surfaces. Usable curves were chosen so that the final reconstructed surfaces were accurate enough for the design of the finger joint prosthesis while remaining applicable from design and data efficiency points of view.

For the joints of different fingers and hands, different dimensions were found even for the left hand and right hand of the same person. The construction of individually adaptable prostheses is not practical from a surgical or economic point of view. This may require in excess of 20 different joint prostheses, which in turn would require a set of surgical tooling for each. Some compromises on the natural joint anatomy are necessary to reduce the number of joint prostheses required (Ash and Unsworth, 2000). It was concluded that a range of sizes of 7, 8, 9 and 10 mm or 7.5, 8.5 and 9.5 mm maximum proximal phalangeal head diameters for the surface replacement prosthesis would cover 97.6 and 91.5% of the population respectively (Ash and Unsworth, 1996). Thus, a generalized method could be used to make suitable prostheses for widespread use instead of making custom-made prostheses for each patient.

7.3   Technological Supports for the Prosthesis Design

Engineering design is defined broadly to include all activities related to the acts of conception and description of engineered products, systems, processes and services, including comparative analysis of alternatives and selection of a preferred alternative. It reaches into other areas, such as manufacture, use and disposal, to ensure good design decision-making.

Engineering design theory and methodology in mechanical engineering was used in this project. A computer-aided method and reverse engineering technology were chosen for the design of the finger joint prosthesis. The laser scan method was used in data capture of the human finger joint bone, which compares well with other data capture methods such as computed tomography (CT) and magnetic resonance imaging (MRI) in reverse engineering (RE).

7.3.1   Reverse engineering

Reverse engineering is a modelling process from original data (which are often digitized from an object) that results in a concise geometric model exportable to CAD/CAM packages. The data points are digitized from the products, which may have been designed before CAD/CAM existed, produced by other manufacturers or made by hand (without CAD design at all). Nowadays, RE is changing from a tedious manual dimensioning or tracing process to a powerful engineering tool utilizing modern digitizing equipment and CAD/CAM systems. The first step in mechanical RE of a geometric model is data acquisition from a part by using some type of digitizer. The two most commonly used digitizers are optical and mechanical. Considering the material of the finger joint bone, the required accuracy, the bone surface complexity and the speed requirement for this project, laser optical scanning is used for the digitizing process.

Reverse engineering in this study involves the scanning of bone samples to determine the geometric range in which the models should fit into and form the basic surfaces to be used for the implants. A laser scanner was used to capture surface data points to create freeform surfaces of human bone samples. This was used to extract curve sets that could be incorporated into a generic design. Other scanning techniques such as coordinate measuring machines (CMMs) were used to compare the accuracy of the 3D scanner with other digitizing machines. The software Paraform was used for the RE procedure.

7.3.2 Comparison of different imaging techniques

Injuries to the hand often result in subtle clinical findings that necessitate a careful evaluation of the bone, cartilage, ligaments and joint capsule for an appropriate diagnosis to be reached. Available tools include standard radiographs in different views, bone scans, arthrography, CT and MR imaging. Standard radiographs remain the most valuable and readily available imaging technique. However, advanced imaging techniques can add significant information about bone, cartilage and soft tissue disorder affecting the hand and wrist, and may influence treatment.

Li (Li, 1999) conducted a study to compare the accuracy of CT and MRI, obtaining the dimensions of finger joint bones using CT, MR imaging and Vernier calipers. It was concluded that CT is the best way to image the hand bones. MRI is adequate for measuring the transverse external dimensions of the metacarpals and phalanges but suffers from not being able to measure the inner cortical dimensions and cross-sectional areas. Radiographs suffer from not being able to measure the inner cortical dimensions and cross-sectional areas. To design good implants for fixation of hand fractures and a good artificial joint, using CT imaging to obtain information about the external dimensions and internal dimensions of hand bones in a large sample is probably the best approach.

For CT, different slice thicknesses are used for different structures. For normal CT, 1 mm or less is used for high resolution of the inner ear, and slices between 5 and 10 mm are used for the chest and for retroperitoneal structures. The resolution of our Minolta 3D digitizer is $200 \times 200 \times 256$ points. Comparing CT with 3D laser scanning, the laser scan method is more accurate. Take a typical proximal phalanx sample as an example, with a volume of $1968\ mm^3$ and a surface area of $1164\ mm^2$. If a 1 mm slice thickness is used and the pixel dimensions are $0.65\ mm^2$, the final CT data are about 37 slices. For the whole surface of the finger joint, we can obtain about 1700 points to describe the surface profile of the sample. Using the Minolta 3D scanner, we can get up to 6000 points on the surface of the same finger joint sample. It is clear that CT shows poor accuracy for an object as small as a finger joint. Laser scanning is better to obtain the external dimensions of finger joint bones. To deal with a larger number of finger joint samples, the laser scanning method is also faster.

7.3.3 Engineering and medical aspects

Medical practitioners often have designs for implants and tools that would potentially make their jobs easier. They often lack the knowledge and resources, however, to see these new designs bear fruit. There are many features of mechanical engineering that are appropriate to the development of medical products. These include manufacturing processes, material testing, design processes and technology. Mechanical engineers have the understanding necessary to assist in medical product development.

7.3.4 NURBS design theory

NURBS is an acronym for 'Non-Uniform Rational B-Spline' and is basically an accurate way to define a freeform curve. NURBS is a mathematical tool for the definition of geometric shapes and is used extensively in the CAD industry and more widely in computing for 3D geometry generation and modelling. The NURBS is an industry standard tool for the representation and design of geometry. Some reasons for the use of NURBSs are that they:

- offer one common mathematical form for both standard analytical shapes (e.g. conics) and freeform shapes;
- provide the flexibility to design a large variety of shapes;
- can be evaluated reasonably fast by numerically stable and accurate algorithms;
- are invariant under affine as well as perspective transformations;
- are generalizations of non-rational B-splines and non-rational and rational Bezier curves and surfaces (Piegl and Tiller, 1997).

A NURBS curve is defined by:

$$C(t) = \frac{\sum_{i=0}^{n} N_{i,p}(t)\omega_i P_i}{\sum_{i=0}^{n} N_{i,p}(t)\omega_i}$$

where p is the order of the B-spline basis functions, P_i are control points and the weight ω_i of P_i is the last ordinate of the homogeneous point P_i^{ω}.

7.4   Proposed Methodology

A systematic method for designing generalized artificial finger joint models is presented here. With such generalized models, accurate shapes of human finger joint prostheses can be designed with greater patient comfort, better range of motion and longer use.

7.4.1   Finger joint model preparation

Ten complete hand specimens from cadavers were used in this research. The specimens were immersed into formalin liquid before the cutting process. They were cleaned very carefully, so the cartilage was left on the specimen in good condition (Figure 7.5). The samples come from various resources and there is no detailed record such as age, gender and similar information. However, it is possible to relate the joint surface to left/right hand and also the finger references/positions.

The bone samples need to be preprocessed for laser scanning. Laser scanning is sensitive to surface finishes. Objects that are too shiny will generate many erratic reflections resulting in

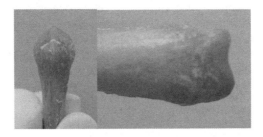

Figure   7.5 Bone samples cleaned with cartilage left intact

Figure 7.6 Sprayed bones used in the project

point clouds full of noise. It is the same for dull surfaces: a flat black surface will absorb too much light and insufficient data will be acquired. Object surfaces with a shiny or dark surface finish are usually sprayed with chalky powders. In this research, the bones were sprayed with SPOTCHECK liquid (SKD-S2 developer), originally used for checking cracks on steels.

Ambient lighting conditions also have to be controlled. For the 3D laser scanner, the better the contrast between the object to be scanned and the background, the better the result will be. The rotary table and the background are both black. Therefore, a fine coating of white SPOTCHECK liquid is preferred to improve the light reflection of the finger joint bone surface (Figure 7.6). It is found that the scanned results can be significantly improved with the bone samples sprayed white rather than sprayed grey or unsprayed.

7.4.2 Finger joint digitization

In this process, the major concerns are the accuracy and efficiency of the captured data. The acquired digital data must be accurate in order exactly to represent the real finger joint geometries, and this is fundamental to the subsequent data manipulations and information analysis. At the same time, the digitalization process must be efficient. There are a large number of bone samples to be inspected in this research to construct generalized models based on statistical analysis. Thus, the digitalizing efficiency may directly affect the overall efficiency of the whole research. Also, the more efficient this process, the greater the economic benefits will be.

Many data acquisition methods, destructive or non-destructive, contact or non-contact, have been developed in the past. These methods are based on different principles, such as optical, electrical, magnetic and mechanical principles or combinations. A general overview of non-destructive data acquisition methods is illustrated in Figure 7.7 (Varady, Martin and Cox, 1997).

We use the Minolta VIVID 700 laser scanner (Figure 7.8), applying the triangulation non-contact optical method as defined in Figure 7.7, to obtain the 3D image data of the finger joint surface.

Laser scanning is a non-contact scanning technique that scans an object by reflecting a laser beam off the object and then detecting it. Employing triangulation-based technologies, 3D coordinate locations on the surface of the object are captured by a sensor on a charge coupled device (CCD) in the probe according to the scan density and pattern parameters set by the user.

In particular, the Minolta VIVID 700 non-contact 3D digitizer was the world's first 3D digitizer with telephoto, making it particularly suitable for scanning small objects. It uses the light-stripe method to emit a horizontal stripe light through a cylindrical lens to the object.

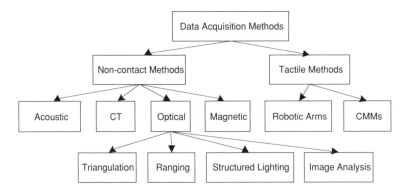

Figure   7.7 Non-destructive data acquisition methods

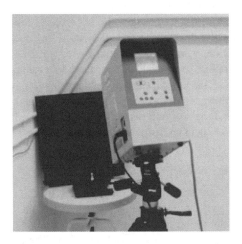

Figure   7.8 Minolta VIVID 700 non-contact 3D digitizer

Figure   7.9 Rotary stage set with bone sample on the centre

The reflected light from the object is received by the 2D CCD array and then converted by triangulation into distance information. This process is repeated by scanning the stripe light vertically on the object surface using a galvano mirror to obtain 3D image data of the object.

Since the unit measuring range of the laser scanner is limited, a rotary stage (Figure 7.9) is used to extend the measurement space. The bone under inspection should be positioned on the approximate centre of the table such that fine adjustment of the 3D image can be facilitated. The rotary stage is rotated by certain angles in each unit inspection, and then these scanned multiple image data are registered into a global model with the aid of the calibration scheme.

The accuracy of the laser scanner is important because we use the 3D point cloud data obtained from the laser scanner throughout the project. Since the reflected light from the object is received by the CCD array and then converted by triangulation into distance information, we assume that this process is linear. However, defects in the lens may result in deformation or non-linearity.

7.4.3   Surface reconstruction in paraform

After data acquisition using the laser scanner, the information gathered is used for further investigation. Reverse engineering techniques are used for the finger joint surface reconstruction. Although RE software Paraform cannot be used directly to perform any useful analysis, its solid modelling tools are quite useful for extracting the surface features and exporting them to other CAD systems for analysis.

7.4.4   Curve feature extraction

After the NURBS surfaces have been reconstructed from the point clouds, the next stage is to extract the geometry features from the surface model. Since only the bearing surfaces are involved in finger joint replacement design, the interesting surface at this stage is the bone surface geometry in the entire proximal phalanx (PP) head and PP/middle phalanx (MP) base portions.

A systematic approach is developed and automated to extract surface features of finger joint models. According to the geometrical features of the human finger joint bones, the head and base portion of the finger joint bones are first identified and isolated from the whole surface. To describe the complex freeform surface of finger joints, the 3D problem is converted to a 2D problem by slicing the 3D models and NURBS fitting on the slicing planes. In this operation, equiangular auxiliary planes are superimposed onto the bearing surface of the finger joint to extract 2D features of the bone surface.

Feature curves are further analysed to find control points that best describe the curves. A database is then constructed to store the information of the control points, and hence to provide a much closer approximation to the actual anatomy of human finger joints.

7.4.5   Database construction and surface generalization

In order to design artificial finger joints that are as close in surface geometry to the original bone joints as possible, real geometries of human finger joint bones are derived using a reverse engineering based method to extract the bone surface features. The database was constructed on the basis of real finger bone information. Statistical analysis is performed on the database to identify the bone surface geometrical features.

Since a custom-made finger joint prosthesis is not practical, a method based on statistical analysis offers a generalized model for finger joint replacement. The sampled bone geometries were categorized into several classes according to the real sizes of the finger joints. The database can be constructed in statistical software such as Microsoft Excel, SPSS, etc.

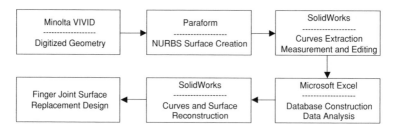

Figure   7.10 Flow chart showing the methodology used

7.4.6   Review of the procedure

The overall methodology in this research can be seen in the flow chart in Figure 7.10.

7.5   Finger Joint Surface Modelling and Feature Extraction

In finger joint digitalization, real finger joint bones are scanned into point clouds, which are used as the source data of the whole process; surface reconstruction is the detailed data manipulation as the medium for subsequent feature extraction; the curve features are the output of this process and will be used in the construction of the generalized model for finger joint replacement design.

7.5.1   Data acquisition of the bone samples

Ten complete hand specimens were used in this research. Figure 7.11 shows the 3D point cloud data obtained from a finger joint surface using the 3D laser scanner. It should be noted that, although the point clouds appear to be continuous, as if it were a surface model, it is not really a continuous surface or solid but many polymeshes tightly packed together.

After data acquisition has been completed, the information gathered is used for further investigation. However, the file format (*.VVD) for the 3D elements from the VIVID system (Figure 7.11) is too specific and cannot be read directly by most solid modelling software. Therefore, the element is exported using the Wavefront format (*.OBJ), which can be imported and read by the reverse engineering software Paraform.

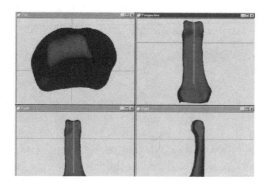

Figure   7.11 3D point clouds of the finger joint bone in VIVID software (shade mode)

The point cloud data in Figure 7.11 are then used for further surface reconstructions. In this process, the output is the reconstructed surface model. The purpose of this operation is explicitly to construct the geometry and topology information from the point clouds. We can then extract the surface curve features from the reconstructed model.

7.5.2   Finger joint surface reconstruction

Since an OBJ file can preserve point cloud, polymesh and even polycurves, all of them are imported into Paraform for analysis (Figure 7.12). Although Paraform could not be used directly to perform any useful analysis, its solid modelling tools are quite useful for extracting the surface features and exporting them to other CAD systems for analysis.

To extract features of the bone surface, a network of curves that define different surface patches need to be created (Figure 7.12), and a series of spring meshes are created in each patch (Figure 7.13). The spring mesh is an intermediate, more uniform polymesh used to generate

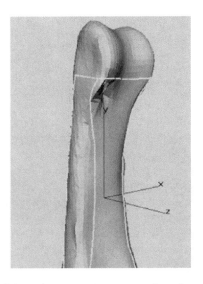

Figure   7.12 Surface patch boundary curves constructed on the surface of the polymeshes

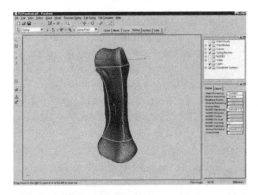

Figure   7.13 Spring meshes created on the polymesh structure

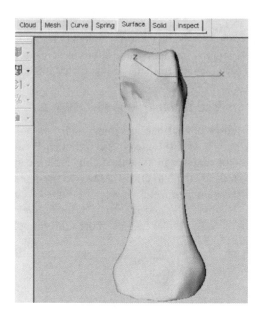

Cloud | Mesh | Curve | Spring | Surface | Solid | Inspect

Figure   7.14 Reconstructed NURBS surface

the NURBS surface. The resolution of the spring mesh governs the maximum resolution of the NURBS surface to be generated. It is necessary to adjust the resolution of the spring mesh according to the prescribed accuracy of the final surface geometry. Trade-offs must be taken between the resolution of the polymesh and the memory/computation consumptions. Excessively dense polymeshes will contribute little to the reconstruction accuracy, and at the same time result in exhausted resource consumptions. Figure 7.14 is a reconstructed surface with a chosen balanced resolution.

The next stage is to export the NURBS surfaces to the CAD software SolidWorks for further manipulation. The NURBS surfaces were exported as IGES files since it is the most commonly used type of file format for handling 3D images between CAD systems.

7.5.3   NURBS curve and feature extraction

After the NURBS surfaces have been reconstructed from the point clouds, the next stage is to extract the geometry features of the surface model. The output is the curves extracted from the reconstructed surface. The purpose of this operation is to analyse the complex shapes of the finger joint surface into lower-level curve features, and then find the control points of such curves and save them into the database for statistical analysis.

Since only the bearing surfaces are involved in finger joint prosthesis design, the interesting surface under study at this stage is the bone surface geometry in the entire PP head and PP/MP base portion.

7.5.3.1   NURBS curve extraction from the PP head

At this point the input surface model contains not only the PP head but also other portions of the bone. One particular problem is therefore how to identify and isolate the PP head from the

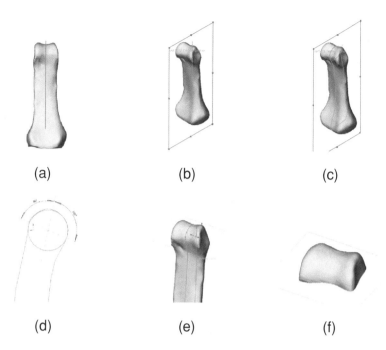

(a)   (b)   (c)

(d)   (e)   (f)

Figure   7.15 Procedure for cutting the head of the PP joint

whole finger joint surface. It is known that the PP head is approximately circular in the sagittal plane. By taking advantage of such unique geometry characteristics of human finger joints, an automatic PP head identification and isolation method is proposed. The detailed algorithm is described as follows:

1. A vertical centre-line is first drawn on the frontal plane (see Figure 7.15). In this plane, the lateral lines are defined as the lines along the lateral surface of the phalanx. Since these are approximately straight lines along the lateral surface, a vertical centre-line can be drawn between them. Because the PP finger joint bone is almost symmetrical in the lateral plane, the centre-line passes through the trough in the frontal plane (see Figure 7.15a).
2. A vertical centre plane is constructed which is parallel to the sagittal plane, and the centre-line is included in the vertical centre plane (see Figure 7.15b).
3. Because the vertical centre plane intersects with the finger joint bone, an intersection curve can be found (see Figure 7.15c).
4. Since the PP head is almost circular in the sagittal plane, the head of the intersection curve is thus of a circular shape. A best-fit circle matched to the PP head in the vertical centre plane is approximated. The centre of the circle is then seen as the centre of rotation (COR) (see Figure 7.15d).
5. The head line is defined as the line along the dorsal surface of the phalanx just proximal to the phalangeal head in the sagittal plane. Since the head line is appropriately a straight line along the dorsal surface, a straight reference line (centre-line) is drawn parallel to the head line through the COR (see Figure 7.15d).
6. The movement of the PIPJs is primarily flexion-extension and is approximately between 0 and 100°. Further investigation of the PP bones also shows that the natural PP head deviates

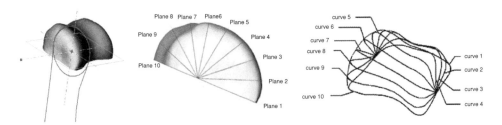

Figure 7.16 NURBS curves extracted from the head of the bone surface

from the circular profile at approximately 100° flexion. In addition, a position up to $45°$ hyperextension of the circular profile is required to allow a $0–100°$ range of movement. In this case, $100°$ flexion to $80°$ hyperextension is used and referenced to the centre-line. This is to imitate the contact range of the real finger joint (see Figure 7.15d).

7. A cutting line is drawn through the centre of rotation with $100°$ flexion to $80°$ hyperextension (see Figure 7.15d).

8. The cutting plane is constructed as follows. An auxiliary line and an auxiliary point are drawn in a plane that is parallel to the top plane and through the COR, where the auxiliary point lies on the auxiliary line. Then, the cutting plane is constructed through the cutting line and the auxiliary point (see Figure 7.15e). The bearing surface of the PP head is then cut using this cutting plane through the cutting line constructed previously (see Figure 7.15f).

After identification and isolation of the PP head from the whole surface, the PP head is superimposed with ten equiangular auxiliary planes over the $180°$ of rotation. A network of intersection curves between the reference planes and the PP head is extracted from the PP head surface. The detailed procedure is as follows (see Figure 7.16):

1. An auxiliary intersection line that goes through the COR and is perpendicular to the vertical centre plane is first constructed in the cutting plane. This auxiliary line is used in the construction of ten equiangular planes and is the intersection line of these planes.

2. The cutting plane is then used as the first plane intersecting with the head of the bone surface. Another nine equiangular planes are then constructed using the auxiliary intersection line. Plane 1 is the cutting plane in Figure 7.16.

3. Since each of the ten equiangular planes intersects with the head of the finger joint bone, ten intersection curves can be found on the surface of the bone head.

In this procedure, only ten equiangular auxiliary planes are superimposed onto the bearing surface of the PP head. However, more planes can be added if more information and high accuracy are needed. This approach is also applicable to the bone shaft or other anatomical models.

The ten extracted curves can be parameterized to obtain the quantitative surface feature of the PP head (Figure 7.17). We will take the fifth intersecting curve as an example:

1. The first dimension that can be made directly is the horizontal distance between the two centre points. The length of the base line is the head width, W_{pph}, of the PP.

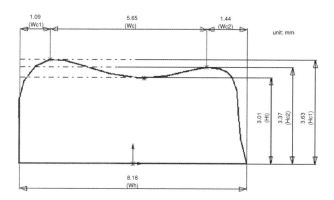

Figure   7.17 Parameterizing the extracted NURBS curves of the PP

2. Three extra reference points are added to describe the trough and crest points. These points are used to dimension their respective vertical heights (H_t, H_{c1}, H_{c2}) from the horizontal base line, which represent their vertical distances from the centre of rotation.
3. The crest points are also used to define the horizontal distance, W_c, based on the respective horizontal distances from the crest points (W_{c1}, W_{c2}). The horizontal distance between the two crests, W_c, is important as it determines the separation between the convex PP head of the bi-condylar hinge joint.

After each of the surface feature curves is parameterized, the control points (crest points and trough point) can be used to construct guide curves which in turn are used to reconstruct the bearing surface of the PP head.

7.5.3.2   NURBS curve feature extraction from the PP and MP base

The curve feature extraction procedure for the base of the PP and MP is similar to that for the PP head. The same problem of identification of the PP/MP base also exists.

A reference plane needs to be constructed to identify and isolate the PP/MP base surface from the whole freeform bone surface. Since the width of the PP/MP base is the key parameter of the PP/MP base, the position of the cutting plane is referenced to the width of the PP/MP base. The cutting plane is horizontal which is different from the cutting plane of the PP head. The distance from the cutting plane to the top of the PP/MP base equals the width of the PP/MP base multiplied by a conversion factor, which is defined as a constant value of 0.25 for the PP/MP base in this research. Figure 7.18 shows an example of the cutting plane for the base of the MP. The width of the MP base is 14.40 mm for this specimen. Therefore, the distance from the cutting plane to the top plane is set to 3.60 mm for this MP model.

The following procedure extracts curve features of the MP base as an example. However, it is also applicable to PP base feature extractions. In fact, the same curve feature extraction method can also be used for the head of the metacarpal phalanx (MCP) bone.

Ten equiangular planes over a 180° angle of rotation are inserted into the head of the MP bone. Each curve can be parameterized to obtain the quantitative surface feature of the PP/MP base. The following procedure takes the MP base as an example. However, it is also applicable to quantitative parameter extraction for the PP base.

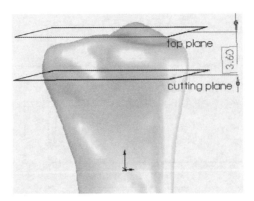

Figure   7.18 Cutting plane inserted into the bone model of the MP

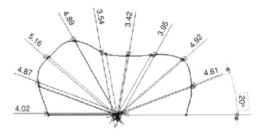

Figure   7.19 Parameterizing the extracted NURBS curves of the MP

Since there are no distinct properties for the extracted surface curves of MP, such as troughs and crests, ten equiangular lines are inserted to divide the curve into several segments (Figure 7.19). These intersection points are saved as the control points of the surface feature and will be used to reconstruct the bearing surface of the PP/MP base.

7.5.3.3   Discussion on curve feature extraction

Different parameterizing methods are used for MP and PP curve extraction. The accuracy of the final reconstructed surfaces varies for different parameterizing methods.

The numbers of reference planes and feature points on each extracted feature curve affect the accuracy of the reconstructed surfaces. Eight analyses were performed on the same MCP bone samples and nine analyses for PP using different numbers of feature curves and control points.

New splines are constructed using the control points, and these are used to loft out the finger joint bone surface. Here, V_0 is the volume of the original bone head, which is compared with the volume of the reconstructed bone head, V_R. The expression

$$\frac{(V_0 \cup V_R) - (V_0 \cap V_R)}{V_0} \times 100\%$$

is used to calculate the volume difference between the original bone head and reconstructed bone head.

For the MCP bone, the number of control points for each spline has more effect on the final reconstructed surface accuracy. The volume difference is less than 2% for 10 control points for each spline. However, with seven and five control points for each spline, the volume difference will not change much. Since the MCP bone head is like a half-sphere, an appropriate number of spline curves and control points can be selected for reconstructing the final surface. Increase in the spline and control point number will not improve the accuracy much.

For the PP bone sample, the parameter of the control points on the spline curves has more effect on the accuracy of the reconstructed surface. For the same number of control points, the final surface accuracy will be greatly improved if the control points are the extreme points of the spline curve. The number of splines and control points also affect the reconstructed surface. Therefore, appropriate positions of control points and an appropriate number of control points and splines should be used according to the required accuracy of the final reconstructed surface.

From the above we can draw the following conclusions: to simplify the feature extraction and ensure accuracy of the final reconstructed surface, appropriate numbers of reference planes and control points should be used for different bone model types.

7.5.4   Automatic surface reconstruction and feature extraction

The above method describes the detailed procedures in finger joint digitalization, surface reconstruction and feature extraction. For each sample, such procedures can be effectively implemented step by step, but, for a large number of finger joint bones, a great deal of data must be analysed and many features extracted. Obviously, it is preferred that such procedures be automated to enhance overall efficiency. An automated implementation of such processes was developed with the application programming interface (API) of the commercial software RapidForm.

RapidForm API is designed for automation. We can access RapidForm objects and reuse codes to develop functions and applications for automation. RapidForm integrates Visual Basic, so we can use macros to automate the curve feature extraction process.

Two macros were developed: one for automated identification of the bearing surface, the other for automated feature extraction.

7.5.4.1   Automated identification of the bearing surface

Automated identification of the bearing surface is implemented using RapidForm API functions. The 3D point cloud data are first imported into RapidForm (Figure 7.20). The macro begins with registration which aligns the scanned shells, followed by a merge operation which combines them into one shell. The final joint model is shown in Figure 7.21. The intersection planes are constructed and corresponding curve features on the surface are derived. Redundant vertices around the lower part of the model are deleted.

7.5.4.2   Automated feature extraction

Two methods were used for the finger joint feature extraction. Method I is called the maximum tangential method (MTM) and method II is called the equiangular method (EM). MTM is suitable for the head of the PP while EM is suitable for the PP and MP bases.

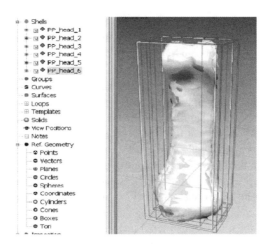

Figure **7.20** Model in .vvd format imported into RapidForm

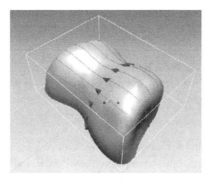

Figure **7.21** Final joint model after the bearing surface identfication process

For the MTM, the operating process can be done either manually or automatically. In this method, we first find two crest points and one trough point as control points, and then other control points are sampled uniformly on the basis of the parameter of each point on the feature curves. To find the extreme points, the feature curve is first tessellated, and then each tessellated point is compared with its neighbouring points. With this method, the extreme points (maximum and minimum point) can be identified.

The EM is almost the same as the scheme described in section 7.5.3. We first find the mutual intersection line of all section planes, which are used to construct intersection rays. Then the middle point of the intersection line is calculated and marked as a centre point. We then construct equiangular rays in each equiangular plane to intersect with the bone model. The intersection points are then saved into the database as the control points.

Using these parameterizing methods, control point information of the feature curves can be extracted and saved into the database. These are fed automatically into statistical software for further analysis, which is extensively used in construction of the generalized model.

7.6   Database Construction and Surface Generalization

As discussed previously, the design for finger joint surface replacement is an effective solution for patients who suffer arthritis disease. One of the critical requirements in the prosthesis is to design artificial joints that are close in surface geometry to the original bone joints so that the best performance of the artificial joints can be obtained.

To achieve this goal, the real geometries of human finger joint bones must be derived first. A reverse engineering based method is employed to extract the surface features of the bones. Since each finger joint bone carries its own unique geometrical features, it is impractical to make custom-made finger joint prostheses from surgical or economical points of view. To resolve this, a novel method based on statistical analysis is proposed to offer a generalized model for finger joint replacement. A finger joint database is constructed for the purpose of data retrieval, and the geometries of the sampled bones are then categorized into several classes according to the real sizes of the finger joints.

7.6.1   Finger joint database construction

In clinical practice, the contact surfaces of the finger joint prove to be critical to the finger joint replacement surgery because the finger joint replacement prosthesis is designed to replace the corresponding damaged bearing surfaces. For this reason, we only focus on the shapes of the finger joints in the head and base section. This principle is applied to all the data samples in the constructed database.

We sampled 79 finger joint bone specimens from ten hands of nine cadavers in total, eight of which are from left-hand joints and the other two (specimens 9 and 6) from right hands. Specimens 6 and 1 are from the same cadaver. For each hand specimen, the MP and PP are extracted from four fingers of each hand excepting the thumb. The head of the PP and the base of the PP and MP were scanned with the Minolta laser scanner.

The following statistical analyses were used to construct the generalized models for the finger joint shapes, based on whole sampled specimens. We analysed the specimens in two directions: first for the statistical *dimension* property, which is used roughly to characterize the general sizes of each finger joint; then statistical *geometry shape* analysis is applied to evaluate the surface geometrical features of finger joint samples.

7.6.1.1   Statistical dimension analysis

Several key parameters of the finger joint dimensions are defined in Table 7.1. The dimensions of the finger joint lengths and widths and the radii of the minimum best-fit circles are extracted. Figure 7.22 showss the general parameters extracted from the PP in the sagittal plane.

The bone length distribution for all samples is shown in Figures 7.23 and 7.24. Figure 7.25 illustrates the PP lengths from individual finger bone specimens.

From Figure 7.23, it is found that the finger joint length distribution is somewhat uneven: only one sample at 38 mm, and there are fewer smaller than larger samples. It can be seen, that although this is quite a good sample size for this kind of project, a larger sample may come up with some different results.

Table 7.1   General parameters and their descriptions extracted from the PP and MP

Parameters	Descriptions
R	Radius of minimum best-fit circle of PP head in sagittal plane
L	Length of PP
L_m	Length from PP base to centre-line
L_p	Length from COR to centre-line
W_{pph}	Width of PP head
W_{ppb}	Width of PP base
W_{mpb}	Width of MP base
Mean	Mean of finger joint dimensions
STDEV	Standard deviation of dimensions

Note: L_p and L_m are measured for an indication of the stem length which is required in the design of the finger joint prosthesis.

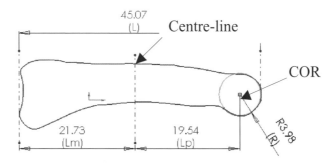

Figure   7.22 PP dimensions in the sagittal plane

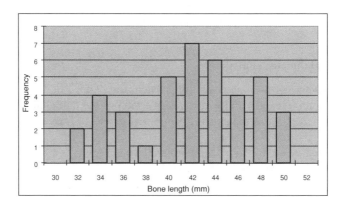

Figure   7.23 Distribution of PP lengths

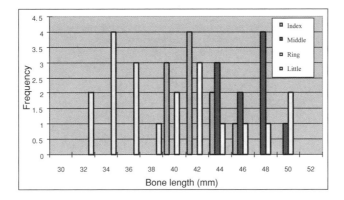

Figure 7.24 Distribution of PP lengths for individual fingers

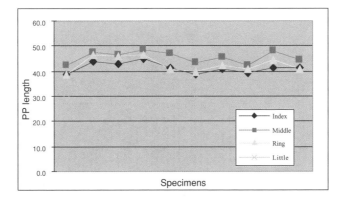

Figure 7.25 PP lengths of individual finger bone specimens

From all these figures, the following statistical assertions can also be made:

- The width distribution trends of the PP head as shown in Figures 7.26 and 7.27.
- The middle finger PP bones tended to be the longest, followed by the ring, the index and then the little finger (see Figure 7.25).
- The index and ring finger tended to have the same PP length, and the mean of the ring finger tended to have a larger length value than the index finger.

Similarly, as in the analysis of the parameters of PP bone length, the following statistical assertions hold true for the parameters of PP head width in the frontal plane:

- The head widths of the PP ranged from 8 to 15 mm and the mean value is 11 mm.
- The middle finger tended to have the widest head in the frontal plane (see Figure 7.28), followed by the index and ring fingers and lastly the little finger bones.
- PP head widths of the index and ring fingers are much closer to each other, and the mean value of the index finger PP head is a little wider than that of the ring finger.

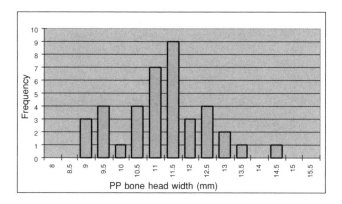

Figure   **7.26** Distribution of PP head widths

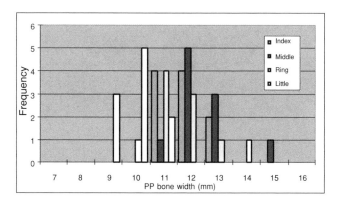

Figure   **7.27** Distribution of PP head widths for individual fingers

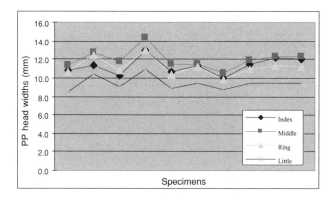

Figure   **7.28** PP head widths of individual finger bone specimens

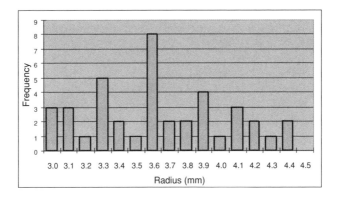

Figure 7.29 Distribution of PP minimum best-fit circle radii in the sagittal plane

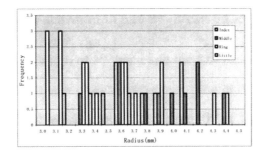

Figure 7.30 Distribution of PP best-fit radii for individual fingers in the sagittal plane

Similar observations were made for the PP base and MP base.

The distribution graphs of the best-fit circle radii are shown in Figures 7.29 and 7.30. The following two statements can be made on the basis of statistical analysis:

- The minimum best-fit radii to the sagittal profiles of the bones ranged from 2.5 to 4.5 mm for the PP head, and the mean value is 3.6 mm.
- The middle finger tended to have the largest best-fit circles followed by the ring, index and little fingers.

The above analysis and statistical statements describe the general dimensions of all sampled data. However, the relationships between these dimensions are still unknown. Therefore, the following analysis is applied to find these relationships.

Relationships between the bone lengths, minimum best-fit bone head radii and bone head/base widths are calculated using correlation coefficients and linear regression.

A scatterplot for widths of the PP head and MP base is shown in Figure 7.31. It can be seen that there is a strong linear relationship between the widths of the PP head and the MP base.

The correlation coefficient R is a linear correlation that reports the strength of the relationship between two dimensions. The correlation coefficients between all the dimensions are shown in Table 7.2. From Table 7.2 it can be concluded that the dimensions are positively correlated.

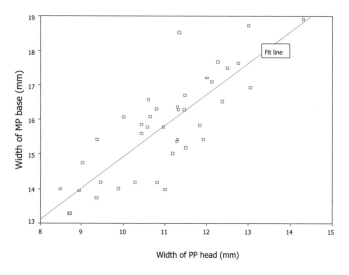

Figure 7.31 Scatterplot for widths of the PP head and the MP base

Table 7.2 Correlation coefficients for R, L, W_{pph}, W_{ppb}, and W_{mpb}

	R	L	W_{pph}	W_{ppb}	W_{mpb}
R	1				
L	0.656	1			
W_{pph}	0.871	0.810	1		
W_{ppb}	0.794	0.607	0.844	1	
W_{mpb}	0.800	0.704	0.916	0.804	1

The correlation coefficient between W_{pph} and W_{mpb} is the largest value among all the correlation coefficients, indicating that the dimensions of the contact surfaces of the PIPJ have a stronger relationship than the other dimensions.

Equations of best-fit lines between any two dimensions using the linear regression method are shown in Table 7.3. The R^2 values and standard errors for each best-fit line are also calculated. The R^2 value is the coefficient of determination, which is a measure of how good the fitted linear relation is. The R^2 values range between 0 and 1. If the R^2 value is close to 1, then the regression line is good. The STE is the standard error of the predicted value in the regression. The linear regression equations can be used to predict the sizes of prostheses required before surgery from X-rays.

7.6.1.2 PP head geometrical features

The finger joint bearing surfaces are analysed in this section. As discussed previously, the requirement for the artificial finger joint geometry and the original finger joint geometry is 'the closer, the better'. We propose two effective methods to model and analyse the shapes from

Table 7.3 Linear regression for dimensions (mm)

Two dimensions	Linear relationship	R^2	STE
L–R	$L = 8.4R + 10.903$	0.433	4.051
L–W_{pph}	$L = 3.234W_{pph} + 5.544$	0.657	3.153
L–W_{ppb}	$L = 2.175W_{ppb} + 6.646$	0.362	4.297
L–W_{mpb}	$L = 2.020W_{mpb} + 14.653$	0.496	3.857
W_{pph}–R	$W_{pph} = 2.786R + 0.982$	0.759	0.662
W_{pph}–W_{ppb}	$W_{pph} = 0.741W_{ppb} - 0.733$	0.669	0.776
W_{mpb}–W_{pph}	$W_{mpb} = 1.272W_{pph} - 0.924$	0.839	0.76
W_{ppb}–W_{mpb}	$W_{ppb} = 0.610W_{mpb} + 7.785$	0.647	0.885
W_{ppb}–R	$W_{ppb} = 2.740R + 5.980$	0.602	0.940
W_{mpb}–R	$W_{mpb} = 3.557R + 0.281$	0.640	1.137

the sampled specimens. We slice each model in two perpendicular directions: first along the sagittal direction, termed the longitudinal direction, and then along the direction that is perpendicular to the sagittal direction, which has been described previously, termed the latitudinal direction.

The PP head has a circular profile in the sagittal plane, varying across its width. The radii of the best-fit circles could be found for the PP to see if the joint is a conforming joint.

Take the PP of the ring finger as an example. Eleven paralleled auxiliary planes are first inserted into the PP head model, with equal distance between each other throughout the width of the PP head. Intersection curves are obtained (Figure 7.32) and fitted into 11 circles in the sagittal plane sections. Since the leftmost and the rightmost curves are not fitted into satisfactory circles according to data fitting verification, they are removed from this analysis (Figure 7.33); the other nine curves are subject to further analysis.

Ideally, the centres of the best-fit circles would be aligned on the same straight line. However, in practice they are not. This can be seen in Figure 7.34. The best-fit circles are obtained using the least-squares fitting method, developed to automate this process.

Data extracted from the best-fit circles are exported to Excel for further analysis. The x, y, z coordinate values and the radii for the best-fit circles with the mean and standard deviations were calculated. Figure 7.35 shows the values for the y, z coordinates and the radii of the best-fit circles from all section curves.

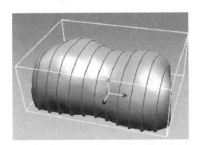 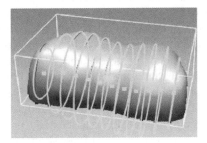

Figure 7.32 Feature curves in the sagittal planes of the PP head (left) and curves fitted to circles (right)

Figure   7.33 Leftmost and rightmost fitted circles in the sagittal plane

Figure   7.34 Centres of best-fit circles in front view and left view

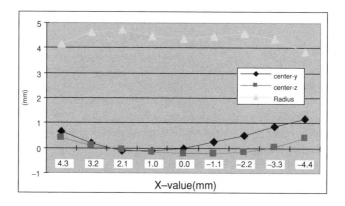

Figure   7.35 Circle profiles in the sagittal plane, varying across the PP head width

From these data we can find that, although the centres of the best-fit circles are not ideally on a straight line, the standard deviation is rather small: for the y, z coordinates the standard deviations are 0.463 and 0.248; for the radius of the best-fit circles the standard deviation is 0.255. The lowest centre point among all these circles tends to be on the left of the PP trough.

Figure 7.36 shows the values for the y, z coordinates and the radii of the best-fit circles for all sampled ring fingers and could be regarded as the generalized model.

It can be found that, for the generalized ring finger models, the circular profiles of the PP head in the sagittal plane vary across its width. The model tends to have the same maximum best-fit circle radius for the two condyles. The centre points are not strictly lying on a straight line. The lowest centre point tends to be coincident with the PP trough, with other centre points varying towards the bearing surface of the proximal interphalangeal joint (PIPJ).

The cartilaginous surface of the MP base is concave and bi-condylar to articulate against the PP head. Tessellated curves are first obtained on the surface of the MP base using two sets of auxiliary planes, perpendicular to each other. The tessellated curves then trim each

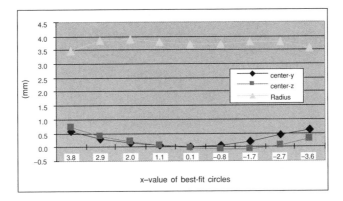

Figure 7.36 Averaged profiles in the plane, varying with PP head width

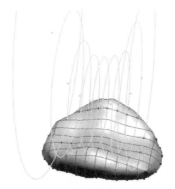

Figure 7.37 Feature curves fitted to circles in sagittal planes for the MP base

other to construct the best-fit circles for the MP base. The trimmed curves are then used to fit corresponding circles in the sagittal plane (Figure 7.37).

From the data on the values of the x, y, z coordinates and the radii for the best-fit circles in the sagittal planes, it was found that:

- The circular profiles do not vary so much across the width of the MP base, except near the edge area.
- The centres of the circles tend to have the same y value, except near the edge area.
- The Z coordinate does not vary so much across the base width.
- The same properties of the circular profiles are also applicable to the averaged circular profiles of the MP base.

The bearing surfaces of the existing finger joint prostheses were designed to be conforming to increase joint stability. However, the above analysis shows that the MP base has a larger radius of curvature than the PP head, indicating that the sides of the joint do not conform to each other.

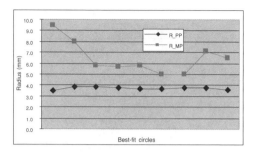

Figure   7.38 Corresponding radii of circular profiles in the sagittal planes for the PP and MP

The averaged radii of the circular profiles in the sagittal planes for the ring fingers of the PP and MP are shown in Figure 7.38. The MP base has a larger radius of curvature than that of the PP head. Therefore, the PIPJ does not match well in essence. However, the radius of curvature does not change much for both phalanges in the middle region of the bearing surfaces. It can be calculated that PP radii of the circular profiles measured from sagittal plane sections are on average 1.67 mm smaller than the MP radii.

Eleven NURBS curves are extracted from the PP head of the finger joint bone in order to find the geometrical features of the bearing surface. The PP head consists of two sagittally circular condyles, which are merged together to form a bi-condylar articular surface and are broader anteriorly than posteriorly. It could be found that the two condyles are not identical, although they are similar in profile shapes and sizes (Figures 7.39 and 7.40).

To analyse the geometrical features of the generalized bone model, finger joint surface feature curves were obtained from all the bone sample specimens. The curve dimensions vary among different curves of the same bone model and the same curve for different bone models. Figure 7.41 shows the fifth curve extracted from all the bone specimens for the PP head. Similar results were also found for the PP base and MP base respectively.

From the above figures we can see that the finger joint surface feature curves follow the same geometrical style for different finger joint specimens. We can categorize the bone samples into several groups in the finger joint surface replacement surgery.

Theoretically, any of the dimensions R, L, W_{pph}, W_{ppb} and W_{mpb} can be used to define the sizes of the finger joint prostheses. From the above analysis it can be found that the PP lengths

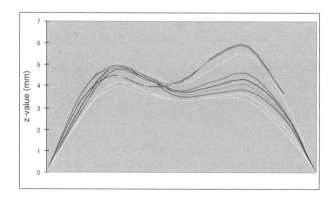

Figure   7.39 Surface curve profiles for the PP head of specimen 2

Figure   7.40 PP head shape: broader anteriorly than posteriorly in the transverse plane

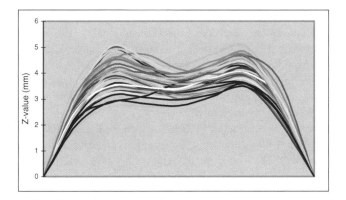

Figure   7.41 PP head bearing surface feature curve profiles of a selected curve from all specimens

have smaller correlation coefficient values compared with other correlation coefficients. Thus, generally, it is better to use other dimensions such as the radii of the minimum best-fit circles and the widths of the PP head as the defining sizes of the finger joint replacement designs.

Suppose that the radii of the best-fit circles in the PP head are used in the finger joint prosthesis design. From the database we find that the radii of the best-fit circles range from 2.5 to 4.5 mm, which is considered as the anatomical range of all finger joints for people ingeneral. Therefore, the final design can be categorized into several groups based on the dimensions of the radii of the best-fit circles.

7.6.2   Generalized finger joint surface reconstruction

Owing to the complexity of the finger joint geometry, a great deal of data is required to construct a model of the articulating surface of a finger joint. Several problems are involved in this process: first, the huge amount of data will require much space to store the information, which is generally not desired; second, there will be strong information redundancy in the database, which will hinder the application of the database. To resolve this, we apply an NURBS-based information model that uses control points and Bernstein polynomials to represent complex curves and surfaces.

Control points on the feature curves of the finger joint surface are extracted and a database stores the control points. Data from the database can then be accessed by commercial CAD software packages to reconstruct the finger joint bearing surfaces.

The bearing surface of each finger joint specimen could be reconstructed using the detailed information obtained from the database. Since it is not practical to provide either a custom-made prosthesis for each joint or the range of sizes for the whole population, we must use a generalized model derived from statistical analysis. Also, it is impossible to have a uniform model for the whole population, so we must further classify the model according to the sizes of human finger joint. To meet the requirements from the whole population, five size classes are suggested after consultation with the designer, S. P. Chow. Therefore, the final design is categorized into five groups according to the radii of the best-fit circles, with a 0.5 mm size increment in the best-fit circle radius size. The radii of 2.5, 3.0, 3.5, 4.0 and 4.5 mm are chosen as the typical geometry sizes.

Although the two condyles of the bi-condylar PP head are not symmetrical (Ash and Unsworth, 1997), the condyles of the reconstructed bearing surfaces are designed to be symmetrical for surgical convenience and economical considerations. The surfaces for the left and right hand are also constructed symmetrical. Therefore, the reconstructed anatomy of the joint is slightly different from the original joint structures; however, this modification will help to reduce the number of joint prostheses required for finger joint replacement surgery. At the same time, we show that such approximations for the original finger joint are acceptable in terms of geometry precision or shape tolerances.

Figure 7.42 shows the reconstructed PP head surface for the middle finger of specimen 5. The difference between the reconstructed surface model and the laser-scanned point cloud data is inspected (Figure 7.43). The inspection deviation analysis is as follows:

- The maximum negative difference is 0.26 mm.
- The maximum positive difference is 0.28 mm.
- The average difference is 0.027 mm and the standard deviation is 0.091 mm.

Figure 7.44 shows the averaged symmetrical PP head surface reconstructed for all PP heads and Chow's finger joint replacement design of the PP head. Chow's design is the latest finger joint prosthesis designed by HKU and still subject to further modifications.

Figure 7.45 shows the reconstructed MP base surface for the index finger of specimen 8. The difference between the CAD reconstructed surface model and the original point cloud data shown is in Figure 7.46.

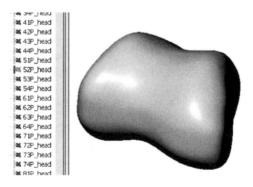

Figure   7.42 PP head surface reconstructed for the middle finger of specimen 5

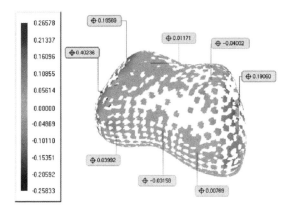

Figure   7.43 Inspection deviation analysis between the CAD model and the original point cloud data for PP head

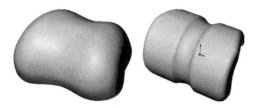

Figure   7.44 Averaged PP head surface reconstructed (left) in comparison with Chow's prosthesis (right)

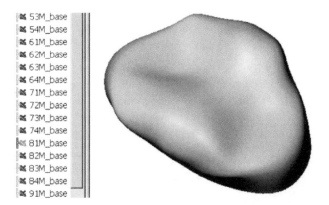

Figure   7.45 Reconstructed MP base surface for the index finger

The inspection deviation analysis results are as follows:

- The maximum negative difference is 0.54 mm.
- The maximum positive difference is 0.52 mm.
- The average difference is 0.014 mm and the standard deviation is 0.14 mm.

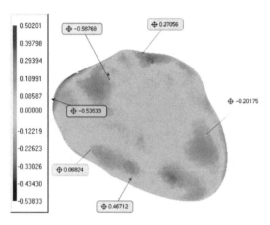

Figure   7.46 Inspection deviation analysis between the CAD model and the original point cloud data for the MP base

Figure   7.47 Averaged MP base surface reconstructed for all MP bases

Figure 7.47 shows the averaged MP base surface reconstructed for all MP bases.

Figure 7.48 shows the reconstructed PP base surface for the ring finger of specimen 3. The difference between the CAD reconstructed surface model and the original laser-scanned point cloud data yielded the following inspection deviation analysis:

- The maximum negative difference is 0.92 mm.
- The maximum positive difference is 0.82 mm.
- The average difference is 0.03 mm and the standard deviation is 0.19 mm.

The above analysis shows that the proposed method is both accurate and flexible in finger joint replacement design. The high precision derived helps to meet the geometrical shape requirements in clinical surgery, and the generalized scheme makes the proposed method practical and economical. To offer a fully customized environment for the finger joint prosthesis design, we employ the 'design table' technique, which is fully parameter driven (Figure 7.49). To modify the sizes in different replacement designs, it is convenient to edit the data relating to the sizes of the patient's finger joint, which is linked to the CAD model and can be automatically updated to match new cases.

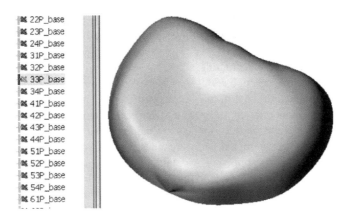

Figure   7.48 Reconstructed PP base surface for the ring finger of specimen 3

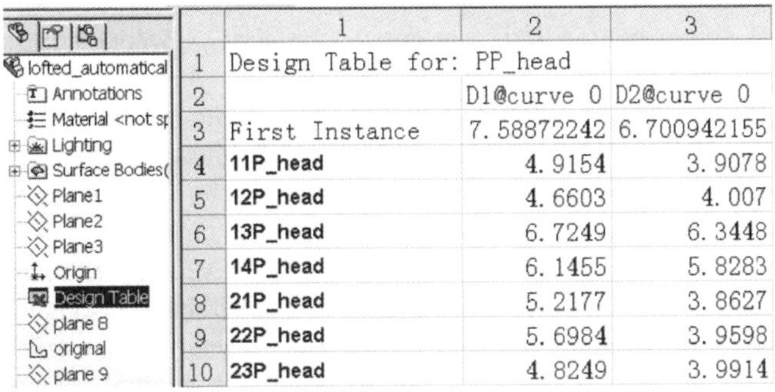

		1	2	3
lofted_automatical	1	Design Table for: PP_head		
Annotations	2		D1@curve 0	D2@curve 0
Material <not sp	3	First Instance	7.58872242	6.700942155
Lighting				
Surface Bodies(	4	11P_head	4.9154	3.9078
Plane1	5	12P_head	4.6603	4.007
Plane2	6	13P_head	6.7249	6.3448
Plane3				
Origin	7	14P_head	6.1455	5.8283
Design Table	8	21P_head	5.2177	3.8627
plane 8	9	22P_head	5.6984	3.9598
original				
plane 9	10	23P_head	4.8249	3.9914

Figure   7.49 Design table used for finger joint prosthesis design

The design table is not only useful in the design process to help verify the best range of sizes for the prostheses but can also be used as a reference for determining what would be the best size and shape of prosthesis to suit a particular finger. In clinical practice, if a patient is suffering from RA and there is a need to perform a finger joint replacement surgery, the patient's data should be registered to the database first. The dimensions of the patient's fingers could be obtained using the CT or MRI method. Not all dimensions of a finger joint need to be extracted since the finger joint dimensions are, in general, linearly correlated. If the head width of a finger joint is used as the reference dimension, the corresponding dimensions of the prosthesis could be found easily in the database. The surgeon can therefore perform the finger joint replacement operation on the patient using the design table designated prosthesis.

7.7   Conclusions

A systematic method for designing the generalized artificial finger joint models for a finger joint prosthesis has been presented here. The generalized finger joint model is generated on statistical information from a large amount of samples of human finger joint specimens. With such generalized models, representative shapes of human finger joints can be designed, which should be able to provide greater patient comfort, a better range of motion and a longer service of life.

A systematic approach to constructing a generalized human finger joint model has been proposed and tested. Reverse engineering based methods, computer-aided design schemes and statistical analyses are utilized in the generalized model construction. In this systematic approach, a 3D laser scanner first digitalizes the specimens of human finger joint samples, and point cloud data are acquired. Surface models are then reconstructed from these point cloud data. By analysing the reconstructed surface model, key curve features are extracted, and corresponding data analysis is applied. The detailed geometry information and characteristic parameters of the curves are then saved in the finger joint database. From statistical analysis of this information, generalized finger joint models can be derived. A number of statistical studies have been carried out using the finger joint database. These have identified different ways to interpret the data collected. It also shows how these data might be useful in assisting with the design of the artificial finger joint and also in the process of matching the joint design to measured patient data.

An automated implementation of the presented approach is developed. Based on the unique geometry characteristics of human finger joints (for example, the minimum radius of the best-fit circle of the PP head), the head and base portions are automatically identified. Auxiliary lines and planes are automatically constructed to retrieve the curve features. Feature-based methods are proposed to detect specific geometry features (for example, the crest points and trough points on the feature curves); key dimensional and geometrical parameters are defined as the characteristic elements of the finger joints. These derived curve features together with the defined parameters are then utilized to construct the generalized model, which statistically represents the model for people in general. The whole process is implemented and executed continuously from one cycle to another. From the original individual case model to the final generalized model, all is done automatically. This significantly shortens the generalized model design process and makes the dynamic database update possible.

The sampled specimens are classified to offer more precise and adaptive generalized models for specific populations. It is not possible to design a fully customized model for each patient. It is also impossible to design a single generalized model for all people from different races and of different age and sex. We propose classifications of the generalized models according to the patient's specific characteristics. As long as the database for the same populations is available, the specific generalized model that best matches the patient can be achieved. Compared with existing finger joint replacement designs, the proposed classified model is more accurate and flexible.

Some of the problems are still not ideally resolved. For example, in this research, only the MP and PP are used in the construction of the finger joint dimensional database. However, MCP information is not taken into account at all. In future work, the MCP can also be extracted and analysed for the surface replacement design of the MCPJ, which is generally believed to be a more complex problem than the PIPJ because of the ligaments surrounding this joint.

In addition, this research is concentrated on the geometrical features of the bearing surfaces of finger joints. Further research on the shaft of the finger joints may be very useful for the design of finger joint prostheses.

Biomechanical aspects of the real human finger joint bone and the finger joint prosthesis should also be investigated. This requires the impact of the joint prosthesis on the human body to be as close to the natural contact forces as possible. Moreover, applications of the proposed method on other human joint structures can also be investigated.

Acknowledgements

The authors would like to acknowledge Prof. S. P. Chow and Mr Terrence Lam for their contribution of radiographic images and advice for this chapter. Part of the work on this project was supported by Hong Kong Research Grants Council grant HKU7290/00M.

References

Ash, H. E. and Unsworth, A. (1996) Proximal interphalangeal joint dimensions for the design of a surface replacement prosthesis. *Proc. Instn Mech. Engrs, Part H: J. Engineering in Medicine*, **210** (H2), 95–108.

Ash, H. E. and Unsworth, A. (1997) Further studies into proximal interphalangeal joint dimensions for the design of a surface replacement prosthesis: medullary cavities and transverse plane shapes. *Proc. Instn Mech. Engrs, Part H: J. Engineering in Medicine*, **211** (H5), 377–90.

Ash, H. E. and Unsworth, A. (2000) Design of a surface replacement prosthesis for the proximal interphalangeal joint. *Proc. Instn Mech. Engrs, Part H: J. Engineering in Medicine*, **214** (H2), 151–63.

Backhouse, K. M. (1968) The mechanics of normal digital control in the hand and an analysis of the ulnar drift of rheumatoid arthritis. *Ann. R. Coll. Surg. Engl.*, **43**, 154–73.

Beckenbaugh, R. D. (1983) Preliminary experience with a noncemented nonconstrained total joint arthroplasty for the metacarpophalangeal joints. *Orthopedics*, **6** (8), 962–5.

Brannon, E. W. and Klein, G. (1959) Experiences with a finger joint prosthesis. *J. Bone and Joint Surg.*, **41A**, 87–102.

Chow, S. P. (2000) Artificial finger joints – its past, present and future. *Chin. J. Hand. Surg.*, **16** (2), 83–5.

Fishback, J. (1999) KUMC Pathology and the University of Kansas. Department of Pathology, University of Kansas Medical Center. http://www.pharmacology2000.com/Hemo/Inflammation/rheumat1_xray.htm

Giurintano, D. J., Hollister, A. M., Buford, W. L., Thompson, D. E. and Myers, L.M. (1995) A virtual five-link model of the thumb. *Med. Eng. Phys.*, **17**, 297–303.

Landsmeer, J. M. (1963) The coordination of finger-joint motions. *J. Bone Joint Surg.*, **45A**, 1654–62.

Li, G. (1999) Magnetic resonance, computed tomographic, radiographic and anatomical correlation of dimensions of the metacarpal and proximal phalanx of the little finger. MPhil thesis, University of Hong Kong.

Linscheid, R. L. and Beckenbaugh, D. J. (1991) Arthroplasty of the metacarpophalangeal joint, in *Joint Replacement Arthroplasty,* Churchill Livingstone, New York, pp. 159–72.

Linscheid, R. L., Murray, P. M, Vidal, M. A. and Beckenbaugh, R. D. (1997) Development of a surface replacement arthroplasty for proximal interphalangeal joints. *J. Hand Surg.*, **22**, 286–98.

Mannerfelt, L. and Andersson, K. (1975) Silastic arthroplasty of the metacarpophalangeal joints in rheumatoid arthritis: long term results. *J. Bone and Joint Surg.*, **57A**, (4), 484–9.

Piegl, L. and Tiller. W. (1997) *The NURBS Book,* Springer, Berlin – New York.

Purves, W. K. and Berme, N. (1980) Resultant finger joint loads in selected activities. *J. Biomed. Eng.*, **2**, 285–9.

Strete, D. (1997) *Pictorial Laboratory Guide for Anatomy and Physiology*, Benjamin Cummings, Menlo Park, CA.

Swanson, A. B. (1969) Finger joint replacement by silicone rubber implants and the concept of implant fixation by encapsulation. *Ann. Rheumatic Diseases*, **28** (5) (supplement), 47–55.

Swanson, A. B. (1972) Flexible implant arthroplasty for arthritic finger joints: rationale, technique and results of treatment. *J. Bone and Joint Surg.*, **54A** (3), 435–575.

Valero-Cuevas, F. J., Zajae, F., Burgher, C. *et al.* (1999) Computer modeling of the hand leads to new concepts in restoring function. Presented at the Fifty-Second Annual Meeting of the American Society for Surgery of the Hand, September, Minneapolis.

Varady, T., Martin, R. R. and Cox, J. (1997) Reverse engineering of geometric models – an introduction. *Computer-Aided Des.*, **29** (4), 255–68.

Zancolli, E. (1968) *Structural and Dynamic Bases of Hand Surgery,* J. B. Lippincott, Philadelphia, PA, pp. 65–75.

8

Scaffold-based Tissue Engineering – Design and Fabrication of Matrices Using Solid Freeform Fabrication Techniques

Dietmar W. Hutmacher

Tissue engineering is an emerging field that allows us to look into a potential future for regenerative medicine. Using this technology, it may become possible to regenerate or replace damaged tissues with laboratory-grown parts such as bone, cartilage, blood vessels and skin. The most common concept underlying tissue engineering is to combine a scaffold/matrix, living cells and/or biologically active molecules to form a 'tissue engineering construct' (TEC) to promote the repair and regeneration of tissues. The scaffold is expected to support cell colonization, migration, growth and differentiation, and to guide the development of the required tissue or to act as a drug delivery device.

Research on the manufacturing of porous scaffold structures for tissue engineering has been carried out for more then three decades. Conventional techniques include solvent casting, fibre bonding and membrane lamination, etc. This work has been reviewed elsewhere. Solid freeform fabrication (SFF) and rapid prototyping (RP) were applied in the 1990s to fabricate complex-shaped scaffolds. Unlike conventional machining, which involves constant removal of materials, SFF is able to build scaffolds by selectively adding materials, layer by layer, as specified by a computer program. Each layer represents the shape of the cross-section of the CAD model at a specific level. Today, SFF is viewed as a high-potential fabrication technology for the

Advanced Manufacturing Technology for Medical Applications   Edited by I. Gibson
© 2006 John Wiley & Sons, Ltd.

generation of scaffold technology platforms. In addition, one of the potential benefits offered by SFF technology is the ability to create parts with highly reproducible architecture and compositional variation across the entire matrix owing to its computer-controlled fabrication. The aim of this chapter is to review the currently applied solid freeform fabrication techniques. In addition, the current challenges and restrictions from a biomaterial point of view are raised and discussed.

8.1 Background

Originally, tissue engineering (TE) was defined from a very broad and general perspective as 'the application of the principles and methods of engineering and life sciences toward the fundamental understanding of structure–function relationships in normal and pathological mammalian tissues and the development of biological substitutes to restore, maintain, or improve functions' (Skalak and Fox, 1988). However, Langer and Vacanti (1993) must be given the credit for having laid down the groundwork for tissue engineering to play a serious role in regenerative medicine (Figure 8.1).

Scaffold-based tissue engineering concepts involve the use of combinations of viable cells, biomolecules and a structural scaffold combined into a 'construct' to promote the repair and regeneration of tissues (Figure 8.2). The construct is intended to support cell migration, growth and differentiation and guide tissue development and organization into a mature and healthy state (Langer and Vacanti, 1993). The science in the field is still young, and different approaches and strategies are under experimental investigation. It is by no means clear what defines ideal scaffold/cell or scaffold/neotissue constructs, even for a specific tissue type. The considerations are complex and include architecture, structural mechanics, surface properties, degradation products and composition of biological components and the changes of these factors with time *in vitro* and/or *in vivo* (Hutmacher, Sittinger and Risbud, 2004; Reece and Patrick, 1998).

Scaffolds in tissue-engineered constructs will have certain minimum requirements for biochemical as well as chemical and physical properties. Scaffolds must provide sufficient initial

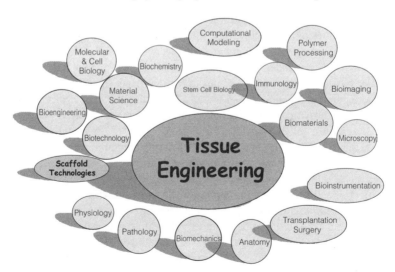

Figure 8.1 Graphical illustration listing the disciplines involved in tissue engineering research and demonstrating the multidisciplinarity

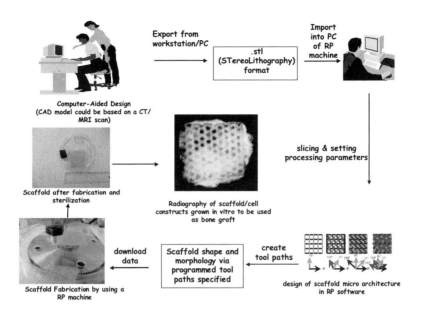

Figure   8.2 SFF/RP technologies allow the development of manufacturing processes that aim to create scaffolds whose macro- and microstructure acts as a template of the extracellular matrix. The flow chart shows how a scaffold/cell construct to be used as a bone graft is created. A flow diagram of the tissue engineering of patient-specific bone grafts by using medical imaging, computational modelling and scaffolds fabricated via RP and made of bioresorbable biomaterials. A computer-generated model (B), which could be based on CT scan data of the patient's bone defect, is imported into the RP system software which allows the model to be 'sliced' into thin horizontal layers, with the toolpath specified for each layer. The 'sliced' data are used to instruct the RP machine to build a scaffold layer by layer, based on the actual data of the computer model. Hollister and Hutmacher's group have shown that different RP technologies are able to produce excellent templates for the treatment of intricate bone defects

mechanical strength and stiffness to substitute for the mechanical function of the diseased or damaged tissue that they aim to repair or regenerate. Scaffolds may not necessarily be required to provide complete mechanical equivalence to healthy tissue, but stiffness and strength should be sufficient at least to support and transmit forces to the host tissue site in the context, e.g. in skin tissue engineering the construct should be able to withstand the wound contraction forces; in the case of bone engineering, external and internal fixation systems might be applied to take the main load bearing until the bone has matured (Hutmacher, 2000).

Cell and tissue remodelling is important for achieving stable biomechanical conditions and vascularization at the host site. Hence, the 3D scaffold/tissue construct should maintain sufficient structural integrity during the *in vitro* and/or *in vivo* growth and remodelling process. The degree of remodelling depends on the tissue itself (e.g. skin 4–6 weeks, bone 4–6 months) and its host anatomy and physiology. Scaffold architecture has to allow for initial cell attachment and subsequent migration into and through the matrix, mass transfer of nutrients and metabolites and provision of sufficient space for development and later remodelling of organized tissue. The degradation and resorption kinetics of the scaffold need to be designed on the basis of two overall strategies described in detail by Hutmacher (2000).

In addition to these essentials of mechanics and geometry, a suitable construct will possess surface properties that are optimized for the attachment and migration of cell types of interest (depending on the targeted tissue). The external size and shape of the construct must also be considered, especially if the construct is customized for an individual patient (Hutmacher, 2000; Sun and Lal, 2002).

In addition to considerations of scaffold performance based on a holistic tissue engineering strategy, practical considerations of manufacture arise. From a clinical point of view, it must be possible to manufacture scaffold/cell constructs under good manufacturing practice (GMP) conditions in a reproducible and quality-controlled fashion at an economic cost and speed. To move the current tissue engineering practices to the next frontier, the manufacturing processes must accommodate the incorporation of cells and/or growth factors during the scaffold fabrication process. To address this, novel manufacturing processes such as robotic assembly and machine- and computer-controlled 3D cell encapsulation are under development so that the tissue-engineered construct not only has a controlled spatial distribution of cells and growth factors but also a versatility of scaffold materials and microstructure within one construct (Figure 8.3).

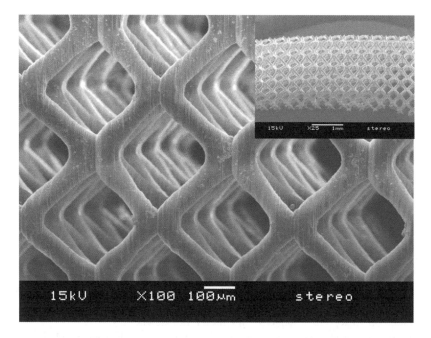

Figure 8.3 SEM images (sample courtesy of Dr Arnaud Bertsch CMI, Switzerland) of a non-degradable scaffold manufactured by micro-SLA. A group from the EPFL Centre of Micro-Nano-Technology has developed a system that can go down to a strut/bar thickness smaller then 50 μm. However, tissue engineers face a limited choice of photopolymerizable biomaterials that have the required biodegradability and mechanical properties. Future work will show if photopolymerizable macromers such as derivatives of PEG acrylate, PEG methacrylate and polyvinyl alcohol (PVA) and modified polysaccharides such as hyaluronic acid and dextran methacrylate can be applied. Poly(propylene fumarate), anhydride and polyethylene oxide (PEO) precursor systems may be explored, as they are already investigated in research or clinical applications typically as curable bioadhesives or injectables

Despite the fact that so-called conventional scaffold fabrication techniques (Ma and Langer, 1999; Hutmacher, 2000; Ma, 2004) have been applied on a large scale by a great number of groups over the last three decades, it must be concluded that most of those matrices are less than ideal for cell-based tissue engineering applications because of deficiencies in mechanical strength, in pore interconnection size and in reproduciblity of the control of variation in porosity and/or morphology within the matrix. In contrast, RP techniques offer unique ways to control precisely the matrix architecture (e.g. size, shape, interconnectivity, branching, geometry and orientation), yielding biomimetic structures of various design and material composition, and to enhance control over scaffold mechanical properties, biologic effects and degradation kinetics. RP techniques are easily automated and integrated with imaging techniques to produce scaffolds that can be customized in overall size and shape, allowing tissue-engineered grafts to be tailored to specific applications or even to individual patients.

Based on this background, it is the aim of this chapter to describe and discuss the state of the art in applying solid freeform fabrication techniques for the design and fabrication of scaffold/cell constructs. The authors also aim to provide an insight into the future direction in this area.

8.2 Introduction

Penning out a road map for the design and fabrication of scaffolds is of fundamental importance to the success of tissue-engineered constructs. Both scaffold chemistry and architecture can influence the fate and function of engrafted cells and neotissue development (Hutmacher, Sittinger and Risbud, 2004). With regard to architecture, macroscopic 3D shapes are typically defined by traditional processes such as extrusion, melt moulding and solvent casting. Material microstructure, in contrast, is often controlled by process parameters such as the choice of solvent in phase separation, the size and morphology of particles for leaching techniques and controlled ice crystal formation and subsequent freeze-drying to create pores. Several research groups have tried to optimize their scaffold fabrication of choice, and this work is reviewed elsewhere (Agrawal, Athanasiou and Heckman, 1997; Ma, 2004; Hutmacher, 2000).

Limitations of conventional scaffold fabrication methods triggered the application of RP/SFF manufacturing techniques, and this trend has grown in popularity over the last decade (Cheah *et al.*, 2004; Sun and Lal, 2002; Hutmacher, Sittinger and Risbud, 2004; Sacholos and Czernuzka, 2003). These techniques offer ways of producing well-controlled, regular microstructures, from a range of suitable scaffold materials, and can be automated and integrated with imaging techniques to produce scaffolds that can be customized both in microstructure and overall size and shape for preparation of implants tailored to specific applications or even to individual patients (Figure 8.3). Some RP/SFF techniques are actually quite slow, whereas others can produce large quantities of formed scaffold quickly and lend themselves to large-scale production of scaffold materials. Most of these techniques offer good to average control over microstructure with pore wall resolutions greater than 100 μm.

8.3 Systems Based on Laser and UV Light Sources

8.3.1 Stereolithography apparatus (SLA)

The most widespread use of SLA in the biomedical industry is presently limited to the creation of accurate models for surgical planning or teaching (Bibb and Sisias, 2002; Brown *et al.*, 2003).

However, there have been research investigations into the fabrication of implantable devices and scaffold using photopolymerizable biomaterials with the SLA method. Theoretically, SLA provides a high degree of control for the manufacturing of the designed scaffold architecture as well as the incorporation of bioagents during the fabrication process.

Stereolithography is based on the use of a focused ultraviolet (UV) laser which is vector scanned over the top of a liquid bath of a photopolymerizable material. The UV laser causes the bath to polymerize where the laser beam strikes the surface of the bath, resulting in the creation of a first solid plastic layer at and just below the surface (Beaman, 1997). The solid layer is then lowered into the bath and the laser-generated polymerization process is repeated for the generation of the next layer, and so on, until a plurality of superimposed layers forming the desired scaffold architecture is obtained. The most recently created layer in each case is always lowered to a position for the creation of the next layer slightly below the surface of the liquid bath. Once the scaffold is complete, the platform rises out of the vat and the excess resin is drained. The scaffold is then removed from the platform, washed of excess resin and then placed in a UV oven for a final curing.

For industrial applications the photopolymer resins are mixtures of simple low molecular weight monomers capable of chain reacting to form solid long-chain polymers when activated by radiant energy within a specific wavelength range. The commercial materials used by SLA equipment are epoxy-based or acrylate-based resins that offer strong, durable and accurate parts/models. However, this material cannot be used as scaffold material owing to lack of biocompatibility and biodegradability. Hence, the limited selection of photopolymerizable biomaterials is a major constraint for the use of the SLA technique in the design and fabrication of scaffolds for tissue engineering applications. However, biocompatible acrylic, anhydride and polyethylene oxide (PEO) based polymers may be explored in future research, as they are already in research or clinical stage typically as curable bioadhesives or injectables. Variation of the laser intensity or traversal speed may be used to vary the crosslink or polymer density within a layer so that the properties of the material can be varied from position to position within the scaffold. This would make it possible to fabricate so-called biphasic or triphasic matrix systems.

In addition, some groups have started innovative work in synthesizing biodegradable and light-curable polymer systems that can be applied to SLA. For example, Cooke et al. (2002) used a custom mixed resin material consisting of poly(propylene fumarate) (PPF), diethyl fumarate (DEF) solvent and bisacylphosphine oxide (BAPO) as the photoinitiator, in a ratio of 140:100:1. The materials were mixed in a standard fume cabinet and attained a similar viscosity as the standard SLA resin. The SLA used was SLA 250/40 from 3D Systems Inc. Cooke aimed to study the regeneration of critical-size defects and designed a Ø50 × 4 mm plate. The part was fabricated and post-cured for 2 h in an UV oven. Cooke described the fabricated part as closely matching the geometry of the CAD design. Features such as holes, slots and protrusions were satisfactorily maintained as the part was built on support structures, but some of the supports (<10%) failed to attach to the build table and the part, resulting in dimples on the underside surface. As a custom-made vat was used, Cooke attributed the manufacturing problems to the lack of adherence of the supports to the build platform, and, as the platform relocated for successive recoats and builds, it disturbed the attachments. The spacing and the diameter of the holes in the build table also affected the flow and spread of the resin. Another suggestion was to minimize resistance of flow and spread of resin as the PPF used was pure and of high molecular weight, with a high viscosity and surface tension. However, making the resin less viscous by adding more solvent would compromise the crosslinking process. Research is currently under way to reach an optimization of the resin viscosity and curing properties. Additionally, a higher degree of crosslinking would decrease the rate of degradation. In conclusion, Cooke's experiment successfully demonstrated the feasibility of using the stereolithography process to build and control 3D multilayer parts made from a biodegradable, biocompatible resin. However, the real challenge for fabricating 3D scaffolds for tissue engineering would be to attain an intricate architecture and high porosity as opposed to a non-porous 2D structure.

Matsuda's group developed a light-curable and biodegradable polymer system based on liquid acrylate end capped poly(caprolactone-*co*-trimethylene carbonate)s [poly(CL/TMC)s] prepared using trimethylene glycol (TMG) or poly(ethylene glycol) (PEG) as an initiator and an acrylate group for subsequent terminal capping, which were used as photocurable copolymers.

The microarchitectured photoconstructs were prepared using a custom-designed apparatus with a moving ultraviolet (UV) light pen driven by a computer-assisted design program. The prepared photoconstructs included microneedles, a microcylinder and microbanks. *In vitro* hydrolytic degradation proceeded with surface erosion when hydrophobic TMG-based photo-cured copolymers were employed, whereas very fast degradation of hydrophilic PEG-based photocured copolymers was observed, probably via concerted actions of surface erosion and bulk degradation. *In vivo* hydrolytic behaviour upon subcutaneous implantation in rats indicated that surface erosion proceeded for TMG-based photoconstructs. Incorporation of an anti-inflammatory drug (indomethacin) into microneedle-structured surfaces showed proof of principle of using this technology for drug delivery applications (Matsuda and Magoshi, 2002; Matsuda and Mizutani, 2002).

Microstereolithography (MSL), in particular, is thought to offer great potential for the production of 3D polymeric structures with micrometre resolution (Figure 8.3). Bertsch, Lorenz and Renaud (1999) described a microstereolithography using dynamic mask generation. The 3D model was sliced to obtain cross-sectional 2D layers with CAD software. To acquire a higher precision for a complex shape, a thousand or more layers have to be generated. Bertsch, Bernhard and Renaud (2001) developed a dynamic pattern generator using a liquid crystal display (LCD) panel. The transparency of the LCD outside the visible wavelength is low, and thus they used an argon ion laser (515 nm wavelength) and controlled the exposure time with a shutter. An optical system was added to redistribute the irradiance and, after the mask, a beam reducer to project it onto the surface of the resin. They reached a resolution of 5 μm. More recently, modifications on the system have been made by different groups, e.g. replacing the laser light source with a broadband Hg light source or using a special LCD compatible with UV. Some systems projected the pattern directly onto the surface of the resin container. Some others focused the light beam on the resin covered with a glass window, forming a constant resin thickness between the window and the substrate. Mauro and Ikuta (2002) described a batch process without a mask. The UV beam is usually scanned over the resin to cause polymerization, but this is time consuming because the UV beam has to run a long way and it is not suitable for mass production. Ikuta used multiple optical fibres to fabricate simultaneously a large number of parts and improve the throughput.

In conventional stereolithography the component layers are built on the surface of the resin. However, two-photon and single-photon microstereolithography have the capacity to build the component directly not only on top of but also inside the resin. Two-photon absorption is an optical non-linear phenomenon that occurs at sufficiently high levels of irradiance when the combined energy of the two photons matches the transition energy between the ground and excited states. The absorption rate is proportional to the square of the incident light intensity but the transition rate is very low, and thus a laser light source of very high power (e.g. several kW) is necessary. To increase as much as possible the non-linear response of the material, the light is focused in a small spot whose volume is as small as possible. Mauro and Ikuta (2002) presented their system of two-photon microfabrication using a mode-locked titanium sapphire laser used to emit a near-infrared (IR) pulse train with a duration of 100 fs and a peak power of 3 kW in the resin. The beam was deflected along the X and Y axes by two scanning mirrors, and the resin container provided the vertical movement. The resin was a mixture of urethane acrylate oligomers/monomers and photoinitiators that was transparent at 770 nm wavelength (the laser wavelength), i.e. it was not polymerized by single-photon absorption. Also, the refractive index was not significantly modified by the solidification, which allowed complex internal structures to be built. Recently, more sensitive photoinitiators have been created, giving more freedom in the choice of resins (Linder et al., 2003).

Mauro and Ikuta (2002) also developed a single-photon process for microstereolithography. A weakly absorbed radiation was chosen to focus tightly on a point inside the resin so that the intensity was sufficiently large to cause polymerization. The resin used was a mixture of urethane acrylate oligomers/monomers and photoinitiators. Compared with the two-photon process, it used an inexpensive He–Cd laser and a simple desktop apparatus. The beam intensity had to be kept sufficiently small to prevent the polymerization of regions outside the focus spot or the resin surface. This system was able to achieve a submicron resolution almost as good as the two-photon process.

8.3.2   Selective laser sintering (SLS)

SLS also uses a focused laser beam, but to sinter areas of a loosely compacted powder. In this method, a thin layer of powder is spread evenly onto a flat surface with a roller mechanism. The powder is then raster scanned with a high-power laser beam. The powder material that is struck by the laser beam is fused, while the other areas of powder remain dissociated. Successive layers of powder are deposited and raster scanned, one on top of another, until an entire part is complete. Each layer is sintered deeply enough to bond it to the preceding layer (Leong, Cheah and Chua, 2003a and 2003b; Paul and Baskaran, 1996).

Rimell and Marquis (2000) have reported the fabrication of clinical implants using a simplified selective laser sintering apparatus and ultrahigh molecular weight polyethylene (UHMWPE). It was observed that solid linear continuous bodies could be fabricated, but material shrinkage occurred when a sheet-like structure was desired. The porosity of the formed material was also a concern. The material exposed to the laser beam was seen to undergo degradation in terms of chain scission, crosslinking and oxidation. It was concluded that the application of this technology to the fabrication of UHMWPE devices requires the development of improved starting powders with increased density.

Lee et al. (1993, 1995) were the first to use SLS to manufacture ceramic bone implants. The augmentation of alveolar ridge defects in canines was conducted to assess the safety and efficacy of these SLS-fabricated calcium phosphate (CaP) implants. Histological evaluation revealed that the implant material was biocompatible and mineralized bone was formed in the macropores. Griffith and Halloran (1996) have reported the fabrication of ceramic parts using suspensions of alumina, silicon nitride and silica particles in UV-photocurable monomer by

SLA. Using a similar technique, a suspension of hydroxyapatite (HA) in photocurable monomer was formulated to produce scaffolds for orbital floor prosthesis. It was concluded that, for bone graft applications, HA scaffolds provide a superior cosmetic appearance compared with conventional techniques. Porter, Pilliar and Grynpas (2001) formulated suspensions of calcium polyphosphate (CPP) and a photocurable monomer for forming bioresorbable skeletal implants using SLA. Sintering CPP at 600°C for 1 h produced a crystalline material (average porosity 22.9 %) exhibiting superior bend strength and toughness compared with amorphous CPP. Tan *et al.* (2003) used different compositions of non-degradable polyether ether ketone (PEEK)/HA powder blends to assess their suitability for SLS processing. However, SLA poses significant material constraints for scaffold fabrication. SLS has the disadvantage that incorporation of sensitive biomolecules is difficult because of the need for local heating of the powder layer so as to sinter it. Nonetheless, low-porosity structures might be built in the future by making use of powders that contain low-melting polymers, such as PEO.

8.3.3   Laminated object manufacturing (LOM)

LOM is a process where individual layers are cut from a laminated sheet (e.g. paper) by a computer-controlled laser, after which the individual layers are bonded together to form a 3D object. LOM has been used for fabrication of bioactive bone implants, using HA and calcium phosphate laminates. The undersurface of the foil has a binder that, when pressed and heated by the roller, causes it to glue to the previous foil. Once the parts have been built, the exterior of the slice is hatched to help the removal of the excess material, as opposed to fluid-based processes (e.g. the SLA process), where the interior is hatched. The disadvantage is the production of burnt edges due to the laser cut, not an issue with most applications, but it creates unwanted and possibly harmful debris in biomedical applications. Material degradation in the heated zone may also occur. The development of a new machine concept makes use of a blade instead of the laser, and this technology might have a greater potential for application in the fabrication of scaffolds.

8.3.4   Solid ground curing (SGC)

Besides the classical laser-based SLA process, alternative processes using digital mask generators, e.g. liquid crystal displays or digital mirror devices (DMDs), have been used successfully to build structures out of polymers and ceramics. In the RP literature this process is also termed solid ground curing (SGC). In contrast to traditional UV laser based SLA machines, DLP systems are significantly cheaper and therefore more versatile with respect to material modifications. At the same time, DLP machines can expose a whole layer at once, whereas laser-based systems have to scan the contour of the object sequentially. DLP systems are based on a digital micromirror device (as used in consumer electronics). By projecting a bitmap onto the photosensitive resin, the liquid resin can be solidified selectively. Theoretically, DLP systems can be used to fabricate scaffolds with high resolution and geometric complexity. However, a prerequisite is the availability of a light-curable, biocompatible and bioresorbable polymer material.

The wider application of SGC in designing scaffolds is mainly driven by developments in photochemically driven gelation technology of biomacromolecules chemically modified with photodimerizable groups. Recent reviews summarize the chemistry and rationale for using these polymers in scaffold-based tissue engineering (Nguyen and West, 2002). Photopolymerizable and biodegradable poly(ethylene glycol)-based macromers, acrylated poly(ethylene glycol) derivatives including poly(ethylene glycol)-*co*-poly(α-hydroxy acid) diacrylate and poly(ethylene glycol)-poly(lysine) diacrylate, both of which are end capped with acryloyl groups, have been studied in detail by Matsuda's group (Mizutani, Arnold and Matsuda, 2002).

Using SCG technology, tubular photoconstructs were prepared by photocopolymerization of vinylated polysaccharide and vinylated gelatin (Mizutani, Arnold and Matsuda, 2002). The mixing of diacrylated poly(ethylene glycol) with vinylated polysaccharide improved the burst strength of photogels against the gradual infusion of water. These photocurable polysaccharides may be used as photocured scaffolds in tissue-engineered devices.

Liu and Bhatia (2002) describe the development of a photopatterning technique that allows localized photoencapsulation of live mammalian cells to control the tissue architecture. Cell viability was characterized using HepG2 cells, a human hepatoma cell line. The utility of this method was demonstrated by photopatterning hydrogels containing live cells in various single-layer structures, patterns of multiple cellular domains in a single 'hybrid' hydrogel layer and patterns of multiple cell types in multiple layers. The authors observed that UV exposure itself did not cause cell death over the doses and timescale studied, while the photoinitiator 2,2-dimethoxy-2-phenylacetophenone was itself cytotoxic in a dose-dependent manner. Furthermore, the combination of UV and photoinitiator was the least biocompatible condition, presumably owing to formation of toxic free radicals.

8.4   Systems Based on Printing Technology

8.4.1   Three-dimensional printing (3DP)

The 3DP technology was developed at the Massachusetts Institute of Technology (MIT) (Cima *et al.*, 1995). 3DP is used to create a solid object by ink-jet printing a binder into selected areas of sequentially deposited layers of powder. Each layer is created by spreading a thin layer of powder over the surface of a powder bed. The powder bed is supported by a piston which descends upon powder spreading and printing of each layer (or, conversely, the ink jets and spreader are raised after printing of each layer and the bed remains stationary). Instructions for each layer are derived directly from a computer-aided design (CAD) representation of the component. The area to be printed is obtained by computing the area of intersection between the desired plane and the CAD representation of the object. The individual sliced segments or layers are joined to form the 3D structure. The unbound powder supports temporarily unconnected portions of the component as the scaffold is built, but is removed after completion of printing.

The solvent drying rate is an important variable in the production of scaffolds by 3DP. Very rapid drying of the solvent tends to cause warping of the printed component. Much, if not all, of the warping can be eliminated by choosing a solvent with a low vapour pressure. Thus, PCL parts prepared by printing chloroform have nearly undetectable amounts of warpage, while large parts made with methylene chloride exhibit significant warpage. It has been found that it is often an advantage to combine solvents to achieve minimal warping and adequate bonding between the biomaterial particles. Thus, an aggressive solvent can be mixed in small proportions with a solvent with lower vapour pressure. After the binder has dried in the powder bed, the finished component can be retrieved and unbound powder removed for post-processing, if necessary (Wu *et al.*, 1996; Giordano *et al.*, 1996; Kim *et al.*, 1998; Park, Wu and Griffith, 1998).

The 3DP process is capable of overcoming the limitations of some SFF techniques in manufacturing certain designs such as overhanging structures. The solution lies in the layering of powders. As the layers are spread, there is always a supporting platform of powder for printing and binding to take place. Thus, as long as the parts are connected together, overhanging structures are of no difficulty. However, one drawback of the powder-supported and powder-filled structure is that the open pores must be able to allow the internal unbound powders to be removed if the part is designed to be porous such as a scaffold for tissue engineering applications. The surface roughness and the aggregation of the powdered materials also affect the efficiency of removal of trapped materials. The resolution of the printer is limited by the specification of the nozzle size and position control of the position controller which defines the print head movement. Another factor is the particle size of the powder used, which simultaneously determines the layer thickness. A layer thickness between 100 and 400 µm can be achieved, depending on the printer.

The versatility of using a powdered material is both an advantage and a constraint of the 3DP process. Most of the available biomaterials do not come in powder form and need special processing conditions to produce a powder that fulfils the requirements for 3DP. The milling of PLA pellets under liquid nitrogen chilled conditions to yield a size of 75–150 µm was a labour intensive and time consuming process as the efficiency in obtaining the microsized powders was low. Cryogenic milling is another method for manufacturing powder from polymeric materials. Despite the restrictions discussed above, 3DP has been explored by several tissue engineering groups for more than a decade.

Cima *et al.* (1995) used an in-house built non-automated 3D printer to fabricate and study scaffold fabrication. The equipment was operated manually, from the raising and lowering of the build platform to the spreading of the powder. A single print nozzle (45 µm) was used in this working prototype. The materials used, poly(ϵ-caprolactone) (PCL) and polyethylene oxide (PEO), were prepared in powder size ranges of 45–75 µm and 75–150 µm. The binder used was a chloroform–PCL solution. The drug release profile or degradation of the device was controlled by printing walls with different thickness and by using the two polymers which have different degradation rates. Through this study (Cima *et al.*, 1995) it was concluded that 3DP could offer several unique build strategies for obtaining zero-order release kinetics, an ideal situation for most drug delivery devices. A highly specific release profile was indeed achievable by reproducible local microstructural control using this SFF.

One of the earlier works on fabricating tissue engineering scaffolds from 3DP used PLGA (85:15) powder packed with salt (NaCl) and a suitable solvent to fabricate scaffolds. The scaffolds were Ø8 × 7 mm in shape with designed interconnected pores of 800 µm and micropores of 45–150 µm resulting from the salt leaching procedure. Overall scaffold porosity was reported to be 60%, a consequence of the salt particles. Kim *et al.* (1998) concluded that the 3DP technique allows the creation of polymer scaffolds with complex macro- and microarchitecture, and that larger sized highly porous devices could be produced compared with the previously limited discs. They have reported on the successful culturing of hepatocytes using the scaffolds in both static and dynamic conditions.

Based on their patent family on 3DP, the MIT group did spin off several companies, among them a company that fabricates scaffolds and drug delivery devices (http://www.therics.com/ Therics Inc.). Zeltinger *et al.* (2001) used scaffolds made by Therics. Poly(L-lactic acid) (PLLA) powder mixed with salt (NaCl) and chloroform as the binder was used. The PLLA was cryogenically milled with liquid nitrogen, after which both the polymer and salt powders were sieved into lots of less than 38, 38–63, 63–106 and 106–150 µm. Disc-shaped scaffolds (Ø10 × 2 mm) were fabricated using two compositions with salt–polymer ratios of 75:25 and 90:10, and the four different lots of particle sizes. Printing was accomplished using a stencil process to control deposition of binder on the polymer powder. After printing the chloroform on the desired regions, the evaporation of the chloroform resulted in precipitation of the polymer around the salt particles. Residual chloroform was extracted from scaffolds by placing them in liquid CO_2 at about 7°C and 800 lbf/in^2 for 5 min. The salt was leached from the scaffolds by immersing in 500 ml of deionized water at room temperature and rotating at 40 r/min for 3 h, changing the water hourly.

Zeltinger et al. (2001) conclude that the TheriForm™ technology could form scaffolds with complex macro- and microarchitectural morphologies. However, they have used a mask (stencil) as in lithography to control part of the printing process. Hence, the microarchitecture was a consequence of the size and morphology of the salt particles and was not dependent on the 3DP process as such. It could be argued that both macro- and microstructures could easily be achieved by using conventional techniques. As only 2 mm thick scaffolds were produced, the critical limitation of salt leaching and solvent dependence in thicker scaffolds was avoided. As a result, the cell culture studies conducted show evidence of the poor interconnectivity of the scaffold morphology produced.

Sherwood et al. (2002) developed an osteochondral scaffold using the TheriForm™ 3D printing process. The material composition, porosity, macroarchitecture, and mechanical properties varied throughout the scaffold structure. The upper cartilage region was 90% porous and composed of PLGA and PLA, with macroscopic staggered channels to facilitate homogeneous cell seeding. The lower, cloverleaf-shaped bone portion was 55% porous and consisted of a PLGA/TCP composite, designed to maximize bone ingrowth while maintaining critical mechanical properties. The transition region between these two sections contained a gradient of materials and porosity to prevent delamination. Chondrocytes preferentially attached to the cartilage portion of the device and biochemical and histological analyses showed that cartilage formed during a 6 week in vitro culture period. The tensile strength of the bone region was similar in magnitude to fresh cancellous human bone. The authors concluded that the fabricated scaffolds had desirable mechanical properties for in vivo applications, including full joint replacement.

Lam et al. (2002) made use of the commercial 3D printer Z402 from Zcorp, one of the six licensees of the 3DP technology from MIT, for their investigations. A number of blends from commercially available biomaterials of natural origin (cornstarch, dextran and gelatin) were used and cryogenically milled into powder (particle size around 100 µm). Distilled water was used as the binder. Different scaffolds designs were produced by CAD software, with different pore sizes and interconnectivities. After fabrication on the 3DP, the scaffolds were post-processed to enhance the strength and increase the resistance against water. The scaffolds were dried at 100°C, after which the unbound powders were removed. A series of infiltration and post-processing methods by using aliphatic polyesters were examined (Figure 8.4). Lam concluded that the method of infiltration with 6.9 ml of PLLA–PCL (75:25) copolymer solution, drying, then immersing for 10 min in water and finally drying again at 100°C produced scaffolds with the best physical properties (Lam, 2000/2001).

The use of natural biomaterials along with water as the binder eliminated the problem of creating a toxic manufacturing environment such as when an organic solvent is used. However,

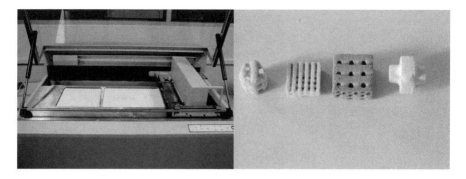

Figure 8.4 3D printing has been utilized by a number of groups to fabricate scaffolds and/or drug delivery devices. The support gained from the powder bed means that overhangs, undercuts and internal volumes can be created as long as there is a hole for the loose powder to escape. Lam *et al.* (2002) made use of the commercial 3D printer Z402 from Zcorp (left image), one of the six licensees of the 3DP technology from MIT, for their investigations. A series of infiltration and post-processing methods by using aliphatic polyesters were examined (image right)

this also created the predicament that the scaffold, now bound by water, was also water soluble. This led to a lengthy post-processing route to 'waterproof' the product. Lam *et al.* (2002) have also pointed out that, as they are all made from bonding powdered particles, the fusion of particles usually would not totally eliminate all the microspaces and gaps in-between the particles, introducing another dimension to the scaffolds of the 3DP process. They have shown that the scaffolds made by the 3DP process possess two types of porosity: the designed interconnected macroporosity; a microporosity caused by the binding process owing to the gaps between the fused particles. This microporosity is likely to be created at random and might be a critical factor in the degradation kinetics of such scaffolds.

Roy *et al.* (2003) studied the *in vivo* performance of porous sintered hydroxyapatite (HA) bone repair scaffolds fabricated using the TheriForm^{TM} solid freeform fabrication process. Porous HA scaffolds with engineered macroscopic channels had a significantly higher percentage of new bone area compared with porous HA scaffolds without channels in a rabbit calvarial defect model at an 8 week time point. Compared with composite scaffolds of 80% polylactic-co-glycolic acid and 20% β-tricalcium phosphate with the same macroscopic architecture as evaluated in a previous study, the porous HA scaffolds with channels had a significantly higher percentage of new bone area. The authors conclude that this study indicates that scaffold geometry, as determined by the fabrication process, can enhance the ability of a ceramic material to accelerate healing of calvarial defects.

Bao *et al.* (2002) developed a simple imprinting technique that allows patterning over a non-flat substrate without the need for planarization. In this process, a polymer film is spin coated onto the mould and then transferred to a patterned substrate by imprinting. By selecting polymers with different mechanical properties, either suspended structures over wide gaps or supported patterns on raised features of the substrate can be obtained with high uniformity. Multilayer 3D polymer structures have also been successfully fabricated using this new imprinting method. The yield and dimensional stability in the multilayer structure can both be improved when polymers with progressively lower T_g are used for different layers. The authors conclude that, compared with existing techniques for patterning on non-flat substrates, the

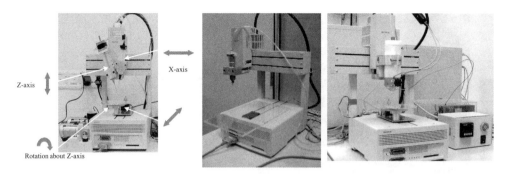

Figure 8.5 A number of groups have developed SFF machines that can perform extrusion of strands/filaments and/or plotting of dots in 3D. Techniques such as FDM, three-dimensional plotting, MJS and PEM share the same basic idea in that a material is extruded in a layered fashion to build a scaffold. Depending on the type of machine, a variety of biomaterials can be used for scaffold fabrication. In this method an extrusion/dispenser head (multiple heads are also possible) is controlled by a three-axis platform, typically an x, y, z table or robot. The images here show a system that was developed at the National University of Singapore. This system can work with a wide variety of polymer hot melts (set-up image left side) as well as pastes/slurries of polymers and ceramics (middle) and reactive oligomers (right)

current method has a number of advantages, including simplicity, versatility, high resolution and low pattern distortion.

8.5   Systems Based on Extrusion/Direct Writing

A number of groups (Sun and Lal, 2002; Sachlos *et al.*, 2003; Hutmacher, Sittinger and Risbud, 2004) and companies (http://www.envisiontec.de, http://www.Sciperio.com) have developed SFF machines that can perform extrusion of strands/filaments and/or plotting of dots in 3D (Figure 8.5). These systems are built to make use of a wide variety of polymer hot melts as well as pastes/slurries (Calvert *et al.*, 1998; Cesarano and Calvert, 2000; Calvert and Crockett 1997), solutions (Ang *et al.*, 2002) and dispersions (Pfister *et al.*, 2004) of polymers and reactive oligomers. Techniques such as fused deposition modelling (FDM), three-dimensional plotting, multiphase jet solidification (MJS) (Koch *et al.*, 1998) and precise extrusion manufacturing (PEM) (Xiong *et al.*, 2001) employ extrusion of a material in a layered fashion to build a scaffold. Depending on the type of machine, a variety of biomaterials can be used for scaffold fabrication.

In the material science literature, another term for extrusion-based systems is used, namely direct writing techniques. Lewis *et al.* assert that direct writing techniques rely on the formulation of colloidal[1] inks for a given deposition scheme. The techniques employed in

---

[1] The 'colloid' is used to describe particles that possess at least one dimension in the size range 10–1 mm. A distinguishing feature of all colloidal systems is that the contact area between particles and the dispersing medium is large. As a result, interparticle forces strongly influence suspension behaviour. Long-range van der Waals forces are ubiquitous, and must be balanced by Coulombic or other repulsive forces to engineer the desired degree of colloidal stability.

direct writing are pertinent to many other fields next to scaffold fabrication such as the capability of controlling small volumes of liquid accurately. Direct writing techniques involving colloidal ink can be divided into two approaches: the droplet-based approach, including direct ink-jet printing and hot-melt printing, and the continuous (or filamentary) technique.

A rapid prototyping (RP) machine was designed and built in house at the Freiburg Materials Research Centre (Albert Ludwigs University, Freiburg, Germany). In contrast to traditional RP systems such as fused deposition modelling, 3D printing, stereolithography and selective laser sintering, which mainly focus on a single mode of material processing, this system was designed to accommodate a much larger variety of synthetic and/or natural biomaterials. Landers *et al.* (2000, 2002) used the term 'bioplotter' to describe the fabrication of scaffolds of different composition. Versatility of the technique was demonstrated in a number of studies. 3D printing (3DP) and 3D bioplotting have been compared in the manufacture of biodegradable polyurethane scaffolds using aliphatic polyurethanes based on lysine ethyl ester diisocyanate and isophorone diisocyanate (Pfiser *et al.*, 2004). Layer-by-layer construction of the scaffolds was performed by 3DP, i.e. bonding together starch particles followed by infiltration and partial crosslinking of starch with lysine ethyl ester diisocyanate. Alternatively, the 3D bioplotting process permitted 3D dispensing and reactive processing of oligoetherurethanes derived from isophorone diisocyanate, oligoethylene oxide and glycerol.

The bioplotter at the Freiburg Materials Research Centre was used to build scaffolds made of PCL, PCL–PEG and PCL–PEG–PCL (Huang *et al.*, 2004). A lay-down pattern of 0/90 was used to form the honeycomb patterns of a square with a single fill gap (FG) of 1 mm. For all three polymers, porous sheets measuring $40 \times 40 \times 4$ mm were fabricated. The cartridge temperature was set at $70°C$ for PCL-PEG, $80°C$ for PCL–PEG–PCL and $100°C$ for PCL, and a 0.5 mm diameter nozzle was used. The scaffolds were built on a piece of white printer paper tapped down on the building platform which was maintained at $20°C$ during the fabrication process. PCL-PEG and PCL-PEG-PCL were dispensed with an air pressure of 4.5 bar instead of 5 bar for PCL. For all three polymers, a speed of 50 mm/min in the x/y axis and a speed of 30 mm/min in the z axis were applied. *In vitro* cell culture studies were conducted by using primary human and rat bone marrow derived stromal cells (hMSC, rMSC). Light, scanning electron and confocal laser microscopy as well as immunocytochemistry studies showed cell attachment, proliferation and extracellular matrix production on the surface as well as inside the scaffold architecture.

Woodfield *et al.* (2002) originally used an in-house built sytem and then later the commercial bioplotter system (Woodfield *et al.*, 2004) for melt extrusion of 3D poly(ethylene glycol) terephthalate–poly(butylene terephthalate) (PEGT/PBT) block copolymer scaffolds with a 100% interconnecting pore network for engineering of articular cartilage. By varying PEGT/PBT composition, porosity and pore geometry, 3D deposited scaffolds were produced with a range of mechanical properties.

A direct writing system that works on the same principles as the 3D bioplotter was first described by Ang *et al.* (2002). The machine is based on an extrusion/dispenser head (multiple heads are also possible) by a three-axis robot. The process generates a scaffold from a computer file (STL, etc.) by building microstrands or dots. Depending on the machine setup, a tissue engineer can make use of a wide variety of polymer pastes/solutions and hot melts as well as dispersions and chemical reactive systems (e.g. fibrin glue). In preliminary studies, chitosan and chitosan–HA scaffolds were produced. For this purpose, solutions of chitosan or chitosan–HA were extruded into a sodium hydroxide and ethanol bath. It was noted that the concentration of sodium hydroxide controlled the adhesion between layers.

Recently, the group fabricated scaffolds from a number of novel di- and triblock copolymers from Michel Vert's group (Vert *et al.*, 1992). Physical properties are under characterization. The cell culture and small animal studies are part of a greater research thrust.

A traditional FDM machine consists of a head heated liquefier attached to a carriage moving in the horizontal x–y plane. The function of the liquefier is to heat and pump the filament material through a nozzle to fabricate the scaffold following a programmed path which is based on a CAD model and the slice parameters. Once a layer is built, the platform moves down one step in the z direction to deposit the next layer. Parts are made layer by layer, with the layer thickness varying in proportion to the nozzle diameter chosen. FDM is restricted to the use of thermoplastic materials with good melt viscosity properties; cells or other theromosensitive biological agents cannot be encapsulated into the scaffold matrix during the fabrication process.

An interdisciplinary group in Singapore has studied and patented the parameters to process PCL and several polymer-based composites (PCL/HA, PCL/TCP, etc.) by FDM (Hutmacher *et al.*, 2001). These first-generation scaffolds (PCL) have been studied for more than 5 years *in vitro* (Figure 8.6) and *in vivo*. The encouraging clinical data allowed the interdisciplinary team to commercialize the technology platform (http://www.osteopore-intl.com). The second-generation scaffolds for bone engineering using FDM were made of polymer and CaP composites as they confer favourable mechanical and biochemical properties, including strength via the ceramic phase, toughness and plasticity via the polymer phase, favourable degradation and resorption kinetics and graded mechanical stiffness. Other advantages include improvement of cell seeding and the enhanced incorporation and immobilization of growth factors. Endres *et al.* (2003) and Rai *et al.* (2004) have tested these PCL/CaP composite scaffolds for bone engineering and reported encouraging results (Figure 8.7).

A variation of the FDM process, the so-called precision extruding deposition (PED) system, was developed at Drexel University and tested (Wang *et al.*, 2004). The major difference

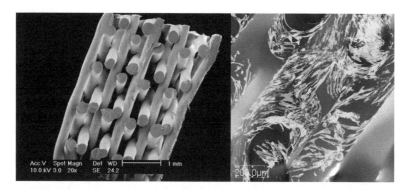

Figure 8.6 SEM image (left) of a PCL scaffold manufactured by FDM using a 0, 60 and 120 lay-down pattern and 70% porosity. The internal scaffold architecture (note that the scaffold was cut with a razor blade after exposure to liquid nitrogen) reveals 100% interconnectivity and a honeycomb-like pore architecture. Confocal laser microscopy in the same plan shows that bone marrow derived precursor cells attach viable cells (stained green), while nuclei of non-viable cells were stained red. Increases in cell densities and high proportions of viable cells were observed over culture periods of 4–6 weeks

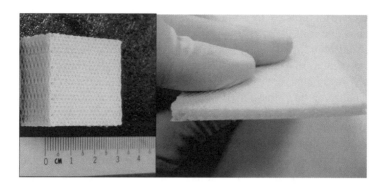

Figure 8.7 Hutmacher *et al.* had been able to evaluate the parameters to process their so-called second generation of scaffolds for bone engineering based on PCL/TCP (Rai *et al.*, 2005) and PCL/HA (Endres *et al.*, 2003) by FDM. They reported that FDM had made it possible to design and fabricate bioresorbable 3D scaffolds with a fully interconnected pore network. Owing to the computer-controlled processing, the scaffold fabrication was highly reproducible. The images here show scaffolds of different size and shape. It should be noted that melt extrusion processes such as FDM employ the continuous deposition of melt material and cannot be stopped during processing. As a result, pore morphology on the edges of a scaffold in both the x and y directions are partially occluded by the deposited material that bridges consecutive roads

between PED and conventional FDM is that the scaffolding material can be directly deposited without filament preparation. Pellet-formed PCL is fused by a liquefier temperature provided by two heating bands and respective thermocouples and is then extruded by the pressure created by a turning precision screw. One such technique, MJS, involves the extrusion of a melted material through a nozzle, which has a jet-like design. The MJS process is normally used to produce metallic or ceramic parts via a lost-wax method. Poly(D, L)-lactide structures for bone and cartilage tissue engineering have been fabricated using an MJS machine built by the Frauenhofer Institute, Stuttgart. The scaffolds had a pore size of 300–400 µm and supported ingrowth of human bone tissues. Calvert *et al.* (1998) built an in-house extrusion-based system that could fabricate scaffolds with a resolution of about 0.5 mm and typical layer heights of 0.2–1.0 mm. Xiong *et al.* (2001) have developed an RP machine termed PEM and fabricated composite scaffolds of poly[L-lactic acid] (PLLA) with tricalcium phosphate (TCP) for bone tissue engineering. A design limitation when using an extrusion system in combination with thermoplastic polymers is the fact that the pore openings for the scaffolds are not consistent in all three dimensions as observed in Figures 8.5 and 8.7. The pore openings facing the z direction are formed in-between the intercrossing of material struts/bars and are determined by user-defined parameter settings. However, for pore openings facing both the x and y directions, these openings are formed from voids created by the stacking of material layers, and hence their sizes are restricted to the bar/strut thickness (diameter). As such, systems with a single extrusion head/liquefier do not vary in pore morphology in all three axes. One design variability exists by extruding one strut/bar directly on top of another.

In another study, the Vozzi group has achieved resolution as low as 10 µm on a 2D structure through the use of a system built by a computer-controlled, three-axis micropositioner, microsyringes and electronically regulated air pressure valves (Vozzi *et al.*, 2003; Vozzi *et al.*, 2002; Ciardelli *et al.*, 2004). However, it needs to be noted that organic solvents also cause cell death, and therefore this process, by using aliphatic polyesters, does not allow the incorporation of growth factors and cells.

Nanodispensing is an alternative method to deposit ultrasmall amounts of liquids on substrate surfaces. The key feature of nanodispensing is deposition of liquids through apertured probe tips commonly used in scanning force microscopy (SFM). Upon contact of the tip and the substrate, liquid at the end of the tip is transferred to the substrate surface. Moving the sample during contact makes it possible to write features with sizes well below 100 nm. This technique is novel and has recently been demonstrated in some laboratories (http://www.csem.ch).

In conclusion, despite some limitations addressed above, the advantage of newer extrusion or direct writing systems is their versatility. However, current encapsulation (Mironov *et al.*, 2003; Desai, 2002) and organ printing techniques that use this SFF technique are restricted to micrometre scale and may not yet permit fabrication of a true 3D tissue construct.

8.6   Indirect SFF

There are two principal routes for the fabrication of scaffolds by RP: indirect and direct routes. Indirect routes rely on an additional moulding step after fabricating the master pattern by RP. The term 'indirect' SFF was coined for scaffold fabrication by Hollister's group (Chu *et al.*, 2002; Hollister, Maddox and Taboas, 2002; Hollister *et al.*, 2000; Hollister, *et al.*, 2001; Taboas *et al.*, 2003). His group used the lost mould technique, combining the epoxy resin moulds made by SLA (SL 5170 and SLA 250/40 from 3D Systems respectively), based on 3D scaffold designs generated from computer-aided design (CAD) software or other imaging techniques, and a thermally curable HA–acrylate suspension as the raw or slurry material. After the moulds were formed, the HA suspension was subsequently cast into the epoxy mould and cured at 85°C. The cured part was placed in a furnace at high temperature simultaneously to burn out the mould and the acrylate binder. Following the mould removal, the HA green body in the designed 3D structure was sintered at 1350°C into a 3D HA scaffold.

In another study by this group, three HA scaffold designs were created with interconnecting pores, one for mechanical evaluation and two for *in vivo* tests. However, the final fabricated scaffolds contained several fabrication inaccuracies. Originally, square pores resulted in bullet-shaped patterns owing to the SLA limitations. Also, after sintering, dimensional changes resulted in changes of up to about 40% in the vertical height of the channels, while other errors were minimal. The *in vivo* experiment demonstrated osteoconductivity and biocompatibility of the HA scaffolds in a minipig model, with up to 16 and 45% bone coverage after 5 and 9 weeks respectively. They have also concluded that the SLA had a limited resolution of 150 × 320 μm curing effectiveness.

Tricalcium phosphate (TCP) scaffolds were fabricated with a suspension of TCP in diacrylate crosslinking monomers using a mould prepared by ink-jet printing. Scaffolds were removed by selective dissolution of the mould. They were heat treated for removal of the acrylic binder, followed by sintering. Considerable linear shrinkage resulted in compromised scaffold porosity after sintering. Composite scaffolds were fabricated from TCP in poly(ethylene glycol) diacrylate using an identical gel casting route (Limpanuphap and Derby, 2002).

In their work published in 2003, Taboas *et al.* (2003) applied the indirect SFF method with conventional sponge scaffold fabrication procedures in creating a series of biomimetic scaffolds for multitissue and structural tissue interface engineering. They designed Ø8 × 8 mm scaffolds with a porosity of 50%. The first step was to create the moulds for the scaffold, this was achieved using wax and polysulphonamide (PSA), which were commercial materials for the Model Maker II 3D printer from SolidScape Inc. The versatility of this machine was that both wax and PSA could be used as the modelling material and/or the support material as there was a choice to remove either material by their respective means. With this setup, four moulds were made. PSA moulds were made by melting and subsequently dissolving the wax portion in Bioact$^{®}$. Wax moulds were obtained by dissolving PSA in acetone, while cement and ceramic moulds were created by casting cement paste in a wax mould and a HA–acrylic slurry in an inverse mould, respectively. The final ceramic (HA) mould was further burnt out and sintered.

With the moulds made, four different casting routes were used to create differently featured scaffolds. The first casting route used three sets of solvent casting. Porogen leaching was used to create a scaffold with local porosity. These solvent casting sets were the traditional salt leaching with salt (104–124 μm) and a PLA (7.5%)–solvent combination, the emulsion–solvent diffusion method using a PLA–tetrahydrofuran and ethanol combination and the snap freezing technique. The second casting route was simply the solvent casting of PLA (25%) into the mould melt. The third casting route involved the melt casting of PGA and PLA into a top–bottom composite scaffold. The fourth casting route produced a polymer–ceramic (HA–PLA) composite by melt casting and an etching process. All the final scaffolds were retrieved by the respective appropriate mould removal process involving melting, dissolving or etching.

Taboas et al. (2003) reported that the ceramic moulds shrank by 50% in volume and that this could be compensated for before the casting, but resulting accuracy needs to be proven in future studies. The 3DP process generated grooved spacings. Generally, any imperfections or crack in the moulds were reflected in the final scaffold. The melt cast scaffolds reveal small inclusions due to trapped air. They have successfully fabricated scaffolds with global interconnected pores ranging from 500 to 800 μm resulting from the prefabricated mould, and when local pores were created they ranged from 10 to 300 μm depending on the local pore creation method.

Wilson et al. (2004) describes the production and characterization of calcium phosphate scaffolds with defined and reproducible porous macroarchitectures and their preliminary in vitro and in vivo bone tissue engineered response. Fugitive wax moulds were designed and produced using a rapid prototyping technique. An aqueous hydroxyapatite slurry was cast in these moulds. After sintering at 1250°C and cleaning, dimensional and material characterizations of the scaffolds were performed. The resulting scaffolds represented the design, and their dimensions were remarkably consistent. A texture inherent to the layer-by-layer production of the mould was impressed onto the vertical surfaces of the scaffolds. Material analyses revealed a α-TCP phase in addition to hydroxyapatite for the moulded ceramics. Non-moulded control ceramics exhibited only hydroxyapatite. Thirty scaffolds were seeded with culture expanded goat bone-marrow stromal cells (BMSCs) and implanted subcutaneously in nude mice for 4–6 weeks. Histology revealed mineralized bone formation in all the scaffolds for both implantation periods. After 4 weeks, bone was present primarily as a layer on scaffold surfaces. After 6 weeks, the surface bone formation was accompanied with bone budding from the surface and occasional bridging of pores. This budding and bridging bone formation was almost always associated with textured scaffold surfaces. However, the area percentage of bone in pores was similar for the 4 and 6 week implantation periods.

The manufacture of collagen-based scaffolds by using the SFF technique to fabricate a mould has also been reported (Sachlos *et al.*, 2003). The mould was dissolved away with ethanol and the collagen scaffold was then critical point dried with liquid CO_2. Another group (Sodian *et al.*, 2002) used SLA models derived from X-ray computer tomography to generate biocompatible and biodegradable heart valve scaffolds from poly-4-hydroxybutyrate (P4HB) and polyhydroxyoctanoate (PHOH) by thermal processing. Use of the thermoplastic elastomers P4HB and PHOH allowed moulding of a complete trileaflet heart valve scaffold without the need for suturing or other post-processing, and demonstrated the ability of indirect SFF to reproduce complex anatomical structures.

SolidScape Model Maker (SMM) uses two heads which deposit a proprietary build and support material. Since both types of material can easily be washed away with appropriate solvents (alcohol and BioAct), this system could be potentially used for SFF techniques. The resolution of the fabricated parts is quite good. The main drawback of the system is the fact that complex cellular structures with severe overhangs require extremely long build times. So far, no report of this system in the tissue engineering literature has been be found.

In conclusion, indirect SFF adds further versatility and detail to scaffold design and fabrication. The previous restriction on casting was the inability of moulds to produce complex geometry and internal architecture. Now, with indirect SFF, traditional casting processes with these SFF moulds can meet the specific tissue engineering requirements, including mechanical integrity and customized shapes. Some highlighted advantages of indirect SFF include cost savings as the materials required for mould casting are substantially less and need not be processed into a dedicated form for any particular SFF process, such as processing into a powder for SLS and 3DP. In addition, indirect SFF allows the usage of a wide range of materials or a combination of materials (composites or copolymers). However, some drawbacks still revolve around this method, including the resolution of the SFF method, as the cast model would inherit the errors and defects from the mould, such as cracks and dimensional changes. Also, a mould removal method must be developed to remove the mould while preserving the casted scaffold intact and the desired properties undisturbed.

8.7   Robotic and Mechatronically Controlled Systems

The main challenge in preparing a useful TEC is to obtain a homogeneous distribution of cells, and hence new tissue, throughout the entire 3D scaffold volume. The morphology of some biodegradable polymer scaffolds, owing to insufficient interconnected macroporosity, has been shown to limit cell colonization, and thus new tissue formation, to the superficial pore layer, resulting in tissue only $\approx$220 µm thick. There exist two possibilities for incorporating cells into the scaffolds: (a) seeding of cells onto the surface of the scaffold subsequent to scaffold fabrication and (b) the incorporation of cells into the scaffold fabrication process. This second approach is of interest, especially when incorporating cells into the scaffold material. For example, special cell encapsulation technologies were introduced (Mironov *et al.*, 2003; see also the above laser-based systems). It should be noted that encapsulation is restricted to the micrometre scale and does not yet permit fabrication of centimetre-scaled scaffolds. Hence, systems based on robotic assembly might offer new possibilities to control.

A robotic microassembly technique has been developed by a multidisciplinary team from the National University of Singapore (Zhang *et al.*, 2002, 2003). The principle of microassembling a functional tissue engineering scaffold is based on the same concept as assembling a structure using small building block units like Lego®. Building blocks of different designs would be first fabricated via lithography or microfabrication technologies assembled by a dedicated precision robot with four-degrees-of-freedom microgripping capabilities, accomplishing a functional-sized scaffold with the required material, chemical and physical properties. A monolithic shape memory alloy microgripper was used to manipulate and assemble the unit microparts into a scaffold structure.

Presently, two designs have been explored. The first is a 2D planar micropart, $420 \times 420 \times 60 \ \mu m3$ in size, and could be fabricated at a precision of 2–3 μm using plasma etching (oxygen plasma), UV LIGA or injection moulding. The second design is a 3D cross-shaped building block with an overall dimension of $500 \times 500 \times 200 \ \mu m$ and a wall thickness of 60 μm. It is axisymmetric and stable in itself. It may be produced using UV LIGA or injection moulding. Both parts may be assembled by pushing a unit part onto another and stacking up a complete scaffold. By applying a suitable force, the microparts would stick together by friction (Figure 8.8).

A group at Carnegie Mellon University, along with its collaborators, is developing a novel technique to materialize the concept of creating vascularized living tissue grafts for direct implantation, using the principles of tissue engineering and SFF. The aim is to infuse cells simultaneously as the scaffold is being synthesized in a layered manner. In order to achieve this, the scaffold fabrication processes must depart from the traditional involvement of heat and toxic chemicals. The overview of this concept entails the fabrication of layered scaffold in a customized geometry derived by the use of clinical imaging data, processing the data and translating the data to the desired scaffold layer ($\sim$1 mm) by a computer numerically controlled (CNC) cutting machine. The material used for the scaffold has been proposed to be a specially formulated polymer–HA composite. The HA would possess varying surface and microstructural attributes embedded *in situ* within a biodegradable polymer backbone. The next step would involve seeding of cells onto each geometrical layer and assembling the cell–scaffold layers together using biodegradable screws, sutures or fasteners. Finally, this combined cell–scaffold construct would be implanted.

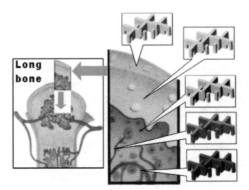

Figure 8.8 A group at the National University of Singapore (Zhang *et al.*, 2002) is developing a novel technique to fabricate scaffold/cell constructs for a variety of tissue engineering applications. The idea is to assemble microscopic Lego®-like building blocks into a scaffold. Based on this CAD-based and automated concept, the distribution of growth factors and living cells within the scaffold can be truly controlled in three dimensions so that scaffold/cell constructs with customized biological and physical properties can be realized

Marra *et al.* (1999) reported the use of the shape deposition manufacturing (SDM) technique to construct osteogenic scaffolds based on blends of PCL and P(D)LGA incorporated with hydroxyapatite (HA) granules for bone tissue engineering applications. The group investigated the use of scaffold fabrication processes which allowed the simultaneous addition of cells to the scaffold interior regions during the scaffold synthesis process. Homopolymer discs as well as blends of 10/90 (10%PCL and 90%PLGA) and 40/60 (40%PCL and 60%PLGA) were prepared with the incorporation of HA ranging between 0 and 50 wt%. The scaffold porosity was reported at 80% by controlling the amount of sodium chloride particles incorporated. They were then cut using a CNC process to generate each cross-section, based on a 3D CAD model. In this manufacturing process the scaffolds were incrementally built up from thin, prefabricated cross-sectional layers of foams (approximately 1 mm thick). To study the feasibility, foam layers were stacked up to form the 3D discs by mating the layers together with biodegradable or non-biodegradable fasteners, including miniature barbs, sutures, screws and nuts.

For the *in vivo* experiments (rabbit model), discs measuring Ø12 × 1 mm were manually assembled using sutures. Each prefabricated layer was first seeded with cells and growth factors before final assembly. However, it was highly dependent on the quality of the prefabricated foams. The salt particle size used to create the micropores and the resultant scaffold pore size had not been reported. Furthermore, the assembly stage of the fabrication process has to be carried out such that there is sufficient pore-to-pore interconnectivity, otherwise the seeded cells would be trapped within each layer, inaccessible to the external supply of nutrients. After 8 weeks, histological analyses showed that the cell-seeded construct appeared as a whole graft while the control showed discernable discrete layers. The group reported bone formation in the seeded implants when compared with the controls (www-2.cs.cmu.edu/People/tissue/front_page.html).

Soft lithography can be used to fabricate a master mould usually made out of polydimethylsiloxane (PDMS). PDMS is chosen because it is chemically inert, non-hygroscopic, isotropic and homogeneous, and has a low interfacial free energy and good thermal stability, durability and ventilation. Vacanti's group (Borenstein *et al.*, 2002) realized channels with a width of about 40 μm and depth of 370 μm using polymer moulding in a silicon mould fabricated by the DeepRIE technique, a plasma etching technology originally patented by Bosch. The goal of this work was to produce organ templates having a feature resolution of 1, well in excess of that necessary to fashion the capillaries comprising the microcirculation of the organ. Initial efforts have resulted in high-resolution polymer scaffolds produced by replica moulding from silicon micromachined template wafers. These scaffolds have been successfully seeded with endothelial cells in channels with dimensions as small as the blood capillaries.

Another process consists of first coating a SU-8 photoresist layer on a silicon wafer. SU-8 is then photopatterned by UV lithography and developed to be a master mould. Then the wafer is coated with PDMS precursor. The whole wafer is degassed in vacuum to eliminate bubbles. This step may be repeated several times so as to remove bubbles completely. Finally, the PDMS precursor solution is cured by baking. After cooling to room temperature, the PDMS mould can be peeled off the substrate and be used for stamping or micromoulding. Folch reports the fabrication of a PDMS mould with defined fluidicinset channels. Polyurethane (PU) precursor droplets were dispensed on the inlet. The outlet was connected with a syringe manually to a Luer lock. The PU precursor was sucked into the mould by the syringe. After curing in UV light, the PU layer with pores was separated from the PDMS mould. The PU layers were stacked under a stereomicroscope system with an alignment precision of about 10–25 μm. The

PU precursor was brushed between each of the two layers which glued the sheets together when the PU scaffold was placed under UV. Vacanti and his group created scaffolds by a similar process but using poly(lactide-co-glycolide) (PLGA). Leclerc, Sakai and Jujii (2003) developed biomedical devices using a PDMS mould. They further applied a photosensitive and biodegradable polymer, poly-(q-caprolactone (CL)-DL-lactide (LA)) tetraacrylate, to fabricate scaffold by PDMS stamping (Leclerc *et al.*, 2004). No reports for further characterization of this type of scaffold were found at the time of writing this chapter. However, the author believes that this material will find its application in tissue-engineered constructs sooner or later.

8.8   Conclusions

Scaffolds are of great importance for tissue engineering (TE) because they enable the fabrication of functional living implants out of cells obtained from cell culture. As the scaffolds for tissue engineering will be implanted in the human body, the scaffold materials should be non-antigenic, non-carcinogenic, non-toxic and non-teratogenic and possess high cell/tissue biocompatibility so that they will not trigger any adverse cellular reactions after implantation. Besides material issues, the macro- and microstructural properties of the scaffold are also very important. In general, the scaffolds require individual external shape and well-defined internal structure with interconnected porosity to host most cell types. From a biological point of view, the designed matrix should serve functions, including:

- an immobilization site for transplanted cells;
- formation of a protective space to prevent soft tissue prolapse into the wound bed and allow healing with differentiated tissue;
- directing migration or growth of cells via surface properties of the scaffold;
- directing migration or growth of cells via the release of soluble molecules such as growth factors, hormones and/or cytokines.

For the three applications described above, solid freeform fabricated scaffolds offer at least three advantages over current technologies for processing biomaterials:

- tailored macroscopic shapes;
- well-defined microstructure, which may include multimodal pore size and distribution as well as directionally oriented pores and channels;
- incorporation of growth factors/cells during manufacture in order to provide controlled release of factors at specific sites.

The different RP/SFF techniques provide the tissue engineer with manufacturing processes that are capable of producing scaffolds from a range of biomaterials and to a range of architectures, morphologies and structures within the design tolerances and parameters based on the chosen tissue engineering strategy (road map). Future work has to provide further evidence that some of these techniques offer the right balance of capability and practicality to be suitable for fabrication of materials in sufficient quantity and quality to move holistic tissue engineering technology platforms into the clinical application.

References

Agrawal, C. M., Athanasiou, K. A. and Heckman, J. D. (1997) Biodegradable PLA–PGA polymers for tissue engineering in orthopedics. *Mater. Sci. Forum*, **250**, 115–28.

Ang, T. H. *et al.* (2002) Fabrication of 3D chitosan–hydroxyapatite scaffolds using a robotic dispensing system. *Mater. Sci. Eng. C*, **20**, 35–42.

Bao, L. R., Cheng, X., Huang, X. D. *et al.* (2002) Nanoimprinting over topography and multilayer three-dimensional printing. *J. Vacuum Sci. and Technol. B*, November–December, **20** (6), 2881–6.

Beaman, J. J. (1997) *Solid Freeform Fabrication: A New Direction in Manufacturing* (eds J. J. Beamann, J. W. Barlow, D. L. Bourell, R. H. Crawford, H. L. Marcus and K. P. McAlea), Kluwer, Boston, MA.

Bertsch, A., Bernhard, P. and Renaud, P. (2001) Microstereolithography: concepts and applications. *Proceedings of 8th IEEE International Conference on Emerging Technologies and Factory Automation*, October 15–18, 2001.

Bertsch, A., Lorenz, H. and Renaud, P. (1999) 3D microfabrication by combining microstereolithography and thick resist UV lithography. *Sensors and Actuators A*, **73**, 14–23.

Bibb, R. and Sisias, G. (2002) Bone structure models using stereolithography: a technical note. *Rapid Prototyping J.*, **8**, 25–9.

Borenstein, J. T. *et al.* (2002) Microfabrication technology for vascularized tissue engineering. *Biomedical Microdevices: BioMEMS, and Biomedical Nanotechnol.*, **4**, 167.

Brown, G. A., Firoozbakhsh, K., DeCoster, T. A., Reyna, Jr, J. R. and Moneim, M. (2003) Rapid prototyping: the future of trauma surgery? *J. Bone Joint Surg. Am.*, **85A** (4) (Suppl.), 49–55.

Calvert, P. and Crockett, R. (1997) Chemical solid free-form fabrication: making shapes without moulds. *Chem. Mater.*, **9**, 650–63.

Calvert, P. *et al.* (1998) Mineralization of multilayer hydrogels as a model for mineralization of bone. *Mater. Res. Soc. Symp. Proc.*, **489**.

Cesarano, J. and Calvert, P. (2000) Freeforming objects with low-binder slurry. US Patent 6027326.

Cheah, C. M., Chua, C. K., Leong, K. F., Cheong, C. H. and Naing, M. W. (2004) Automatic algorithm for generating complex polyhedral scaffold structures for tissue engineering. *Tissue Eng.*, March–Aprril, 10 (3–4), 595–610.

Chu, T. M. G., Ortone, D. G., Hollister, S. J., Feinberg, S. E. and Halloran, J. W. (2002) Mechanical and *in vivo* performance of hydroxyapatite implants with controlled architectures. *Biomaterials*, **23**, 1283–93.

Ciardelli, G., Chiono, V., Cristallini, C., Barbani, N., Ahluwalia, A., Vozzi, G., Previti, A., Tantussi, G. and Giusti, P. (2004) Innovative tissue engineering structures through advanced manufacturing technologies. *J. Mater. Sci. Mater. Med.*, April **15** (4), 305–10.

Cima, M., Sachs, E., Fan, T. L., Bredt, J. F., Michaels, S. P., Khanuja, S., Lauder, S., Lee, S. J., Brancazio, D., Curodeau, A. and Tuerck, H. (1995) US Patent 5387380.

Cooke, M. N., Fisher, J. P., Dean, D., Rimnac, C. and Mikos, A. G. (2002) Use of stereolithography to manufacture critical-sized 3D biodegradable scaffolds for bone ingrowth. *J. Biomed. Mater. Res., Part B: Appl. Biomater.*, **64B**, 65–9.

Desai, T. A. (2002) Microfabrication technology for pancreatic cell encapsulation. *Expert Opin. Biol. Ther.*, **2**, 633–46.

Endres, M., Hutmacher, D. W., Schantz, J. T., Kaps, C., Ringe, J., Salgado, A. J., Reis, R. L. and Sittinger, M. (2003) Osteogenic induction of human bone marrow derived mesenchymal progenitor cells in novel synthetic polymer/hydrogel matrices. *Tissue Eng.*, **9** (4), 689–702.

Giordano, R. A., Wu, B. M., Borland, S. W., Cima, L. G., Sachs, E. M. and Cima, M. J. (1996) Mechanical properties of dense polylactic acid structures fabricated by three dimensional printing. *J. Biomater.* (Polymer Science Edition), **8** (1), 63–75.

Griffith, M. L. and Halloran, J. W. (1996) Freeform fabrication of ceramics via stereolithography. *J. Am. Ceram. Soc.*, **79**, 2601–08.

Hollister, S. J., Chu, T.M., Halloran, J.W. and Feinberg, S. E. (2001) Design and manufacture of bone replacement scaffolds, in *Bone Mechanics Handbook* (ed. S. C. Cowan), CRC Press, Boca Raton, FL, Vol. **36**, 1–14.

Hollister, S. J., Levy, R. A., Chu, T. M. G., Halloran, J. W. and Feinberg, S. E. (2000) An image based approach to design and manufacture craniofacial scaffolds, *Int. J. Oral/Maxillofac. Surg.*, **29**, 67–71.

Hollister, S. J., Maddox, R. D. and Taboas, J. M. (2002) Optimal design and fabrication of scaffolds to mimic tissue properties and satisfy biological constraints. *Biomaterials*, October, **23** (20), 4095–103.

Huang, M. H. *et al.* (2004) Degradation and cell culture studies on block copolymers prepared by ring opening polymerization of ε-caprolactone in the presence of poly(ethylene glycol). *J. Biomed. Mater. Res.*, **69A** (3), 417–27.

Hutmacher, D. W. (2000) Polymeric scaffolds in tissue engineering bone and cartilage. *Biomaterials*, **21**, 2529–43.

Hutmacher, D. W., Sittinger, M. and Risbud, M. V. (2004) Scaffold-based tissue engineering: rationale for computer-aided design and solid free-form fabrication systems. *Trends Biotechnol.*, **22** (7), 354–62.

Hutmacher, D. W., Teoh, S. H., Zein, I. and Tan, K. C. (2001) Mechanical properties of polycaprolactone scaffolds fabricated using fused deposition modelling. *J. Biomed. Mater. Res.*, **55**, 1.

Kim, S. S., Utsunomiya, H., Koski, J. A., Wu, B. M., Cima, M. J., Sohn, J., Mukai, K., Griffith, L. G. and Vacanti, J. P. (1998) Survival and function of hepatocytes on a novel three-dimensional synthetic biodegradable polymer scaffold with an intrinsic network of channels. *Ann. Surg.*, **228**, 8–13.

Koch, K. U., Biesinger, B., Arnholz, C. and Jansson, V. (1998) Creating of bio-compatible, high stress resistant and resorbable implants using multiphase jet solidification technology, in *Time-Compression Technologies*. Time-Compression Technologies '98 Conference, London, Rapid News Publications, pp. 209–14.

Lam, C. X. F. (2000/2001) A study on biomaterials for 3-D printing. B.Eng. thesis, Department of Mechanical Engineering, NUS, Singapore.

Lam, C. X. F., Mo, X. M., Teoh, S. H. and Hutmacher, D. W. (2002) Scaffold development using 3D printing with a starch-based polymer. *Mater. Sci. Eng. C – Bio S.*, **20**, 49–56.

Landers, R. and Mülhaupt, R. (2000) Desktop manufacturing of complex objects, prototypes and biomedical scaffolds by means of computer-assisted design combined with computer-guided 3D plotting of polymers and reactive oligomers. *Macromol. Mater. Eng.*, **282**, 17–21.

Landers, R. *et al.* (2002) Fabrication of soft tissue engineering scaffolds by means of rapid prototyping techniques. *J. Mater. Sci.*, **37**, 3107–16.

Landers, R. *et al.* (2002) Rapid prototyping of scaffolds derived from thermoreversible hydrogels and tailored for applications in tissue engineering. *Biomaterials*, **23**, 4437–47.

Langer, R. and Vacanti, J. P. (1993) Tissue engineering. *Science*, **260**, 920–6.

Leclerc, E., Miyata, F., Furukawa, K. S., Ushida, T., Sakai, Y. and Fujii, T. (2004) Effect on liver cells of stepwise microstructures fabricated in a photosensitive biodegradable polymer by softlithography. *Mater. Sci. and Eng. C*, **24** (3), 349–54.

Leclerc, E., Sakai, Y. and Fujii, T. (2003) Cell culture in 3-dimensional microfluidic structure of PDMS (polydimethylsiloxane). *Biomed. Microdevices*, **5** (2), 109–14.

Lee, G., Barlow, J. W. (1993) Selective laser sintering of bioceramic materials for implants. Proceedings of Solid Freeform Fabrication Symposium, August 9–11, 1993, Austin, TX, pp. 376–80.

Lee, J. J., Sachs, E. M. and Cima, M. (1995) Layer position accuracy in powder-based rapid prototyping. *Rapid Prototyping J.*, **1**, 24–37.

Leong, K. F., Cheah, C. M. and Chua, C. K. (2003a) Classification of rapid prototyping systems, in *Rapid Prototyping, Principles and Applications* (eds K. F. Leong, C. M. Cheah and C. K. Chua), World Scientific Publishing, Singapore, pp. 19–23.

Leong, K. F., Cheah, C. M. and Chua, C. K. (2003) Solid freeform fabrication of three-dimensional scaffolds for engineering replacement tissues and organs. Review. *Biomaterials*, June, **24** (13), 2363–78.

Limpanuphap, S. and Derby, B. (2002) Manufacture of biomaterials by a novel printing process. *J. Mater. Sci. – Mater. in Med.*, **13** (12), 1163–6.

Linder, V., Wu, H., Jiang, X. and Whitesides, G. M. (2003) Rapid prototyping of 2D structures with feature sizes larger than 8 microm. *Anal. Chem.*, May 15, **75** (10), 2522–7.

Liu, V. A. and Bhatia, S. N. (2002) Three-dimensional photopatterning of hydrogels containing living cells. *Biomed. Microdevices*, **4**, 4257–66.

Ma, P. X. (2004) *Materials Today*.

Ma, P. X. and Langer, R. (1999) Morphology and mechanical function of long-term *in vitro* engineered cartilage. *J. Biomed. Mater. Res.*, **44**, 217–21.

Marra, K. G., Szem, J. W., Kumta, P. N., DiMilla, P. A. and Weiss, L. E. (1999) *In vitro* analysis of biodegradable polymer blend/hydroxyapatite composites for bone tissue engineering. *J. Biomed. Mater. Res.*, **47**, 324–35.

Matsuda, T. and Magoshi, T. (2002) Preparation of vinylated polysaccharides and photofabrication of tubular scaffolds as potential use in tissue engineering. *Biomacromolecules*, **3**, 942–50.

Matsuda, T. and Mizutani, M. (2002) Liquid acrylate endcapped biodegradable poly(ϵ-caprolactone-co-trimethylene carbonate). II. Computer-aided stereolithographic microarchitectural surface photocon-structs. *J. Biomed. Mater. Res.*, **62**, 395–403.

Mauro, S. and Ikuta K. (2002) Submicron stereolithography for the production of freely movable mechanisms by using single-photon polymerization. *Sensors and Actuators A: Phys.*, 100: 70–76.

Mironov, V. *et al.* (2003) Organ printing: computer-aided jet-based 3D tissue engineering. *Trends Biotechnol.* **21**, 157–61.

Mizutani, M., Arnold, S. C. and Matsuda, T. (2002) Liquid, phenylazide-end-capped copolymers of epsilon-caprolactone and trimethylene carbonate: preparation, photocuring characteristics, and surface layering. *Biomacromolecules*, **3**, 668–75.

Nguyen, K. T. and West, J. L. (2002) Photopolymerizable hydrogels for tissue engineering applications. *Biomaterials*, **23**, 4307–14.

Park, A., Wu, B. and Griffith, L. G. (1998) Integration of surface modification and 3D fabrication techniques to prepare patterned poly(L-lactide) substrates allowing regionally selective cell adhesion. *J. Biomater. Sci. Polym. Ed.*, **9** (2), 89–110.

Paul, B. K. and Baskaran, S. (1996) Issues in fabricating manufacturing tooling using powder-based additive freeform fabrication. *J. Mater. Process. Technol.*, **61**, 168–72.

Pfister, A. *et al.* (2004) Biofunctional rapid prototyping for tissue-engineering applications: 3D bioplotting versus 3D printing. *J. Polym. Sci. Part A: Polym. Chem.*, **42**, 624–38.

Porter, N. L., Pilliar, R. M. and Grynpas, M. D. (2001) Fabrication of porous calcium polyphosphate implants by solid freeform fabrication: a study of processing and *in vitro* degradation characteristics. *J. Biomed. Mater. Res.*, **56**, 504–15.

Rai, B. *et al.* (2005) The effect of rhBMP-2 on canine osteoblasts seeded onto 3D bioactive polycapro-lactone scaffolds. *Biomaterials*, **26** (17), 3739–48.

Reece, G. P. and Patrick Jr, C. W. (1998) Tissue engineered construct design principles, in *Frontiers in Tissue Engineering* (eds C. W. Patrick Jr, A. G. Mikos and L. V. McIntire), Elsevier Science, New York, pp. 166–96.

Rimell, J. T. and Marquis, P. M. (2000) Selective laser sintering of ultra high molecular weight polyethylene for clinical applications. *J. Biomed. Mater. Res.*, **53**, 414–20.

Roy, T. D., Simon, J. L., Ricci, J. L. *et al.* (2003) Performance of hydroxyapatite bone repair scaffolds created via three-dimensional fabrication techniques. *J. Biomed. Mater. Res.*, **67A** (4), 1228–37.

Sacholos, E. and Czernuszka, J. T. (2003) Making tissue engineering scaffold work. Review on the application of solid freeform fabrication technology to the production of tissue engineering scaffolds. *Eur. Cells Mater.*, **5**, 29–40.

Sachlos, E. *et al.* (2003) Novel collagen scaffolds with predefined internal morphology made by solid freeform fabrication. *Biomaterials*, **24**, 1487–97.

Sachs, E. M., Haggerty, J. S., Cima, M. J. and Williams, P. A. (inventors) (1989) US Patent US5204055: Three-dimensional printing techniques, Massachusetts Institute of Technology, Cambridge, MA (applicant), issued/filed dates: April 20, 1993 / December 8, 1989.

Sherwood, J. K., Riley, S. L., Palazzolo, R., Brown, S. C., Monkhouse, D. C., Coates, M., Griffith, L. G., Landeen, L. K. and Ratcliffe, A. (2002) A three-dimensional osteochondral composite scaffold for articular cartilage repair. *Biomaterials*, December, **23** (24), 4739–51.

Skalak, R. and Fox, C. F. (1988) *Tissue Engineering*, Alan R. Liss.

Sodian, R. *et al.* (2002) Application of stereolithography for scaffold fabrication for tissue engineered heart valves. *ASAIO J.*, **48**, 12–16.

Sun, W. and Lal, P. (2002) Recent development on computer aided tissue engineering: a review. *Comput. Meth. Programs Biomed.*, **67**, 85–103.

Taboas, J. M. *et al.* (2003) Indirect solid free form fabrication of local and global porous, biomimetic and composite 3D polymer–ceramic scaffolds. *Biomaterials*, **24**, 181–94.

Tan, K. H. *et al.* (2003) Scaffold development using selective laser sintering of polyetherethcrkctone–hydroxyapatite biocomposite blends. *Biomaterials*, **24**, 3115–23.

Vert, M., Li, S. M., Spenlehauer, G. and Guerin, P. (1992) Bioresorbability and biocompatibility of aliphatic polyesters. *J. Mater. Sci.: Mater. in Med.*, **3**, 432–46.

Vozzi, G., Flaim, C., Ahluwalia, A. and Bhatia, S. (2003) Fabrication of PLGA scaffolds using soft lithography and microsyringe deposition. *Biomaterials*, June, **24** (14),2533–40.

Vozzi, G. *et al.* (2002) Microsyringe-based deposition of two-dimensional and three-dimensional polymer scaffolds with a well-defined geometry for application to tissue engineering. *Tissue Eng.*, **8**, 1089–98.

Wang, F. *et al.* (2004) Precision extruding deposition and characterization of cellular poly-ϵ-caprolactone tissue scaffolds. *Rapid Prototyping J.*, **10** (1), 42–9.

Wilson, C. E. *et al.* (2004) Design and fabrication of standardized hydroxyapatite scaffolds with a defined macro-architecture by rapid prototyping for bone-tissue-engineering research. *J. Biomed. Mater. Res.*, **68**, 123–32.

Woodfield, B. F. *et al.* (2004) Design of porous scaffolds for cartilage tissue engineering using a three-dimensional fiber-deposition technique. *Biomaterials*, **25**, 4149–61.

Woodfield, T. B. *et al.* (2002) Scaffolds for tissue engineering of cartilage. *Crit. Rev. Eukaryot. Gene Expr.*, **12**, 209–36.

Wu, B. M., Borland, S. W., Giordano, R. A., Cima, L. G., Sachs, E. M. and Cima, M. J. (1996) Solid free-form fabrication of drug delivery devices. *J. Controlled Rel.*, **40**, 77–87.

Xiong, Z. *et al.* (2001) The fabrication of porous poly[L-lactic acid] scaffolds for bone tissue engineering via precise extrusion. *Scr. Mater.*, **45**, 773–9.

Zeltinger, J., Sherwood, J. K., Graham, D. A., Mueller, R. and Griffith, L. G. (2001) Effect of pore size and void fraction on cellular adhesion, proliferation and matrix deposition. *Tissue Eng.*, **7**, 557–71.

Zhang, H., Burdet, E., Hutmacher, D. W. and Poo, A. N. (2003) Robotic microassembly of scaffolds for tissue engineering (Video). IEEE International Conference on Robotics and Automation (ICRA' 03), 2003, Taipei, Taiwan.

Zhang, H., Burdet, E., Hutmacher, D. W., Poo, A. N., Bellouard, Y., Clavel, R. and Sidler, T. (2002) Robotic micro-assembly of scaffold/cell constructs with a shape memory alloy gripper. Proceedings of IEEE International Conference on Robotics and Automation (ICRA' 02), 2002, Washington, DC.

9

Direct Fabrication of Custom Orthopedic Implants Using Electron Beam Melting Technology

Ola L. A. Harrysson and Denis R. Cormier

One of the most significant challenges to more widespread usage of custom implants involves their fabrication. The mass production techniques used to produce generic implants by the thousand are not easily scaled down for custom implant production. Custom implants are therefore much more costly and time consuming to produce than their generic counterparts. This chapter discusses the use of the electron beam melting (EBM) process to produce custom biomedical implants more cost effectively. The EBM process is a layered manufacturing technique capable of producing fully dense metal parts starting from metal powder. A 4.8 kW electron beam selectively melts one layer at a time, thus producing freeform objects that require little finish machining. At present, the EBM process can produce steel and titanium components, and new materials are currently under development. This chapter includes examples of custom knee implants and bone plates that were built using the EBM process.

9.1   Introduction

Millions of people worldwide undergo surgery every year to receive orthopedic joint implants or bone plates. The most common types of joint replacement involve hip or knee joints, and the number of surgeries conducted each year is expected to increase substantially in the coming decade. The rise in joint replacement surgeries is attributable to factors such as an increase in the

Advanced Manufacturing Technology for Medical Applications   Edited by I. Gibson
© 2006 John Wiley & Sons, Ltd.

elderly population, more active lifestyles and increasing obesity. Unfortunately, replacement joints do not last indefinitely. When the implant has worn out or one of the components has come loose, the patient must go through a revision surgery. This involves removing the old implant and inserting a new one. Revision surgeries are often much more complicated than the primary surgery, and most people can only go through 1–3 revisions in their life. The longevity of an implant depends on many factors. As a rule of thumb, implants last 10–15 years for older patients and even less for younger and/or more active patients [1]. An important factor is the initial fit of the implant, which is influenced by the size of the implant selected by the physician and the accuracy of the bone preparation. According to a study conducted by Toksvig-Larsen and Rigd [2], the average contact surface between the bone and the implant is only 53% owing to the hand tools and cutting guides used. Most hip and knee implants come in 5–7 generic sizes/shapes and are designed on the basis of the 'average' patient. A large percentage of implant revisions become necessary because of loosening of components. Loosening is often caused by bone remodeling resulting from stress concentrations on the bone structures (i.e. the bone changes shape). The current hand tools used in surgery to shape the bone to fit the prosthesis have limited the design of implant components in the past. Since saws, drills and reamers are used to shape the bone, the interface between the implant and the bone must involve flat or cylindrical surfaces. To solve some of the problems related to bone remodeling and to achieve a better fit for each patient, custom-designed implant components have recently started to appear. In many cases, the same orthopedic hand tools are used in surgery, and the custom implant is shaped the same as the conventional implants. The only difference is that the size of the implant is customized for the patient. In the extreme, researchers can customize the entire implant by designing custom tools and cutting guides to go along with implants in which both the size and shape have been customized to fit the patient. With the introduction of orthopedic robots, the shape constraints have been removed, and a robot can shape the bone to virtually any geometry [3, 4]. The robot uses an end-mill to machine the bone and works much like a six-axis CNC machine.

9.2   Literature Review

In many instances, generic mass-produced implants will not work well with patients having abnormal anatomy. In some cases, the surgeon must reconstruct the joint to fit the implant. A more desirable approach is to custom design implants that compensate for the abnormal anatomy. The resulting surgical procedure is generally far less complicated. The primary barrier to the widespread use of custom-designed implant components is the cost associated with the procedure. However, the literature shows many examples where custom-designed implant components have been used to solve complicated cases.

9.2.1   Custom joint replacement implants

Keenan et al. [5] reported on seven cases where patients suffered from supracondylar fractures above their total knee arthroplasties. This is a very serious complication that can occur immediately after the primary joint replacement or many years down the road. When the distal femur fractures right above the femoral component, it is very difficult for the surgeon to reattach the bone and get a stable fixation. In these seven cases, the surgeons decided to use custom designed long-stem endo-rotating knee prostheses fabricated by WaldemarLink,

GmbH, Hamburg, Germany. All patients successfully recovered and gained acceptable range of motion and mobility.

Sathasivam et al. [6] described five cases where patients were in need of total knee arthroplasty, yet standard implant components would not suffice owing to abnormal geometry of the joint. Constrained condylar knee implants were custom designed for each patient. The first implant that was designed was parameterized so that the same basic design could be used for all patients by changing certain dimensions. The implants were fabricated on a four-axis CNC machine using stainless steel. No surgical complications were encountered, and the custom implants were functioning well in each patient's 1–4 year check-ups.

Joshi et al. [7] report on the results of custom acetabular components that were designed to address acetabular deficiencies on 27 patients. All patients had a history of multiple hip revisions with severe bone deficiency in the acetabular region. Computed tomography (CT) scans were obtained for the patients, and rapid prototyping (RP) models were created of the pelvic areas. Custom acetabular components were designed for each patient, and prototypes were created to test the fit of the component on the RP model of the pelvis. After verifying the design, the custom acetabular components were fabricated in titanium using a CNC machine. Some of the patients experienced complications, but the overall results were promising and custom acetabular components were recommended by the authors for complicated cases.

Hip prostheses are very common, and most implant manufacturers offer a wide range of sizes and models. However, the shape and the size of the medullar canal can be very different from one patient to another. Consequently, several implant companies have started to offer custom-designed hip prostheses. One problem with custom implants is the need for custom tools and guides that are required if an orthopedic robot is not available to do the cutting operations. Werner et al. [8] describe a methodology for custom design of hip prostheses using CAD/CAM technology [8]. The researchers emphasize the importance of collaboration between the designer and the surgeon to achieve the optimal design. A parametric model of a hip implant is used as the starting point, and the exact parameters are derived from a CT scan where the contours of the bone are imported into a CAD program. The stem cannot always be designed fully to fill the medullar cavity since the prosthesis must be able to be inserted without cracking the bone.

Viceconti et al. [9] describe the HIDE computer software that is used for the custom design of hip implants. Instead of converting the patient-specific CT images into a CAD-like model that can be imported into a standard CAD package, they developed software that reads CT images that are used directly for the custom design of the implants. When the design is ready, the program generates the toolpath information for a CNC machine to fabricate the custom-designed hip implant. According to their study, the software is easy to use, and the operators had a short learning curve before they were proficient in designing implant components.

Wolford et al. [10] reported on an evaluation of 42 patients treated with a custom-made total joint prosthesis for the temporo-manibular joint (TMJ). All patients had severe problems that would have been difficult to correct using standard TMJ implants. A CT scan was obtained of each patient, and a full-scale model of the temporo-manibular area was produced using stereolithography. Custom implant components were designed on the basis of the CT data and tested for perfect fit and function on the RP model. Few patients had complications, and all of them greatly increased the function of their TMJ and reduced the pain.

In a collaborative research project between the Department of Industrial Engineering and the College of Veterinary Medicine at North Carolina State University (NCSU), the custom design and fabrication of hip and knee implants are currently being investigated. The proposed concept is to custom design implants based on patient-specific CT data. Since the proposed design of the knee implants involves the use of novel contoured bone–implant interface surfaces, they must be tested on animals before any clinical studies can be conducted on humans. The veterinary school at NCSU is currently performing hip replacements on dogs on a weekly basis using standard implant components. Both a cemented and a cementless hip implant system have been developed at NCSU where the implant components come in several sizes. Currently there are no commercial knee implants available for dogs owing to the number of sizes that would be necessary to accommodate different sizes of dog knees within a breed and between different breeds. The idea of custom designing the implant components for each patient would solve this problem for dogs, and would provide a remedy for the common condition of knee arthritis in dogs. The project started by looking at the custom design of the femoral and the tibial component for a human patient [11]. The idea was to look at the performance of standard implant components and to try to improve the longevity of the prosthesis by optimizing the design. The new implant design calls for the articulating surface of the femoral component to mimic the original shape of the distal femur for each patient (Figure 9.1). This eliminates the need to resurface the patella unless necessary owing to patellar osteoarthritis. It also reduces changes in the patient's walking gait because of altered joint geometry. The bone–implant interface has been designed to even out the stress distribution on the bone to reduce bone remodeling which can cause loosening of the component. The use of a contoured bone–implant interface surface also allows surgeons to minimize the amount of bone that needs to be removed during the surgery. To fit the proposed femoral component design to the bone, the shape of the distal femur must be obtained. Orthopedic robots are capable of producing these freeform

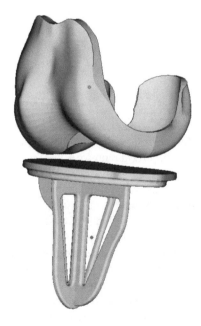

Figure   9.1 Computer model of the proposed human knee implant design

surfaces, although they are not yet approved for human use in all parts of the world. A freeform cutting device is under development at North Carolina State University to enable surgeons to produce a contoured surface on the distal femur that will match that of the custom implant [12]. The tibial component has been designed so that the tray will follow the exact contour of the proximal tibia after the initial cut is completed. This optimizes support from the cortical bone and reduces the risk of having the tray protrude into the cancellous bone. The stem part of the tibial component was designed to be supported by the cortical bone to prevent movement of the component which can lead to loosening.

The same concept has been used to design custom knee implant components for dogs in collaboration with the College of Veterinary Medicine [13, 14]. First, a CT scan of the canine patient's knee is imported into the Mimics software package produced by Materialise. The CT images are edited and converted into a three-dimensional (3D) model of the knee joint using thresholding and region growing commands. The 3D model is exported from Mimics as an STL file and is then converted into NURBS surfaces using the Raindrop Geomagic Studio software package. The 3D NURBS surfaces are imported into CAD software and serve as the basis for the design of the custom components. As shown in Figure 9.2, a parametric computer model has been developed to shape the bone–implant interface for the custom femoral component

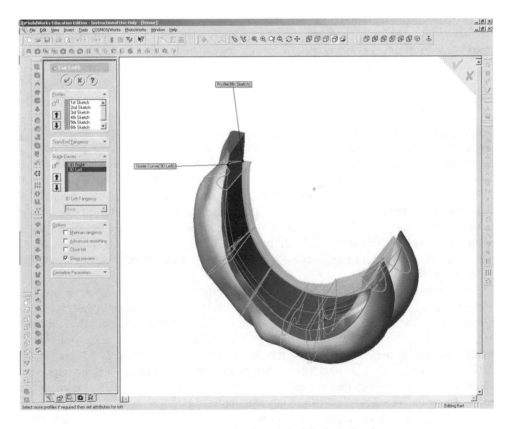

Figure   9.2 Parameterized model of the knee implant

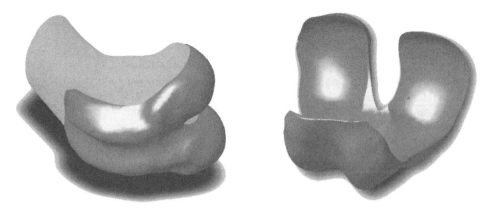

Figure 9.3 Computer model of the custom femoral component

based on the 3D model of the distal femur derived from the CT data [15]. The parametric model substantially reduces the time needed to design a custom implant. The bone–implant interface surface is used to remove the arthritic bone tissue in the model, and the custom implant is designed to replace the removed bone tissue. The model is customized for each by changing the appropriate geometric parameters in the model. In this case, a CT scan of a Labrador Retriever with arthritic knees was used to conduct the experiment and to evaluate the concept. Figure 9.3 shows a rendering of the custom knee implant fitted to the distal femur.

9.2.2   Custom bone plates and implants

While the previous section discussed several studies involving the custom design of joint replacement implants, work has also been done in the area of custom bone plates. Van Steenberghe *et al.* [16] investigated the use of custom-designed drill guides for the place-ment of zygoma implants. Without the drill guides, it is quite difficult to obtain precise hole locations. Although a CT-based computer planning system is available to assist in determining where the holes should be located, transferring the exact location obtained from the software to the patient is difficult. This project used 3D information available from the 3D CT-based planning software to design and fabricate custom drill guides. Researchers in this study re-ported that the placement of the zygoma implants improved significantly when custom drill guides were used.

Eufinger and Saylor [17] described how custom-designed cranial implants were tested on 135 patients at 25 health care centers in Europe between 1993 and 1999. Cranioplasty is usually a very difficult and time-consuming surgery where the implant component has to be fabricated prior to the procedure. If the missing piece of bone is large, it is challenging to design an implant that will fit perfectly and follow the overall curvature of the skull. In this project, CT data of the patients' craniums were used to custom design the implants. All surgeons reported a substantial decrease in both the time and complexity of the surgery. All patients experienced a much shorter recovery time than usual. Only two implants failed owing to wound dehiscence.

Hieu *et al.* [18] reported on a similar study conducted in Asia where patient-specific CT data were used to custom design cranial implants. The goal of this project was to design and fabricate custom cranial implants at the lowest possible cost. To optimize the cost, a mold was fabricated using a three-axis CNC machine, and the implants were cast using polymethyl methacrylate (PMMA). The complex surface of the implant caused difficulties with the CNC machining, and four- or five-axis machining could have solved some of those problems. However, the paper described an inexpensive way of producing custom-designed cranial implants.

Bone plates are frequently used to repair bone fractures or to reattach the bone parts after an osteotomy. These bone plates are often made from titanium and usually come as straight metal plates with predrilled holes for screw attachment. The orthopedic surgeon usually spends a considerable amount of time during the surgical procedure to shape the plate to fit the contour of the patient's bone. The surgeon uses simple hand tools and special bending devices to achieve the correct shape. This procedure can be quite difficult in some cases. After shaping the plate to follow the bone contours, the plate is attached to the bone using titanium screws. The predrilled holes do not always line up with the preferred screw location, and the physician sometimes must insert screws in less than ideal locations. The problem is particularly difficult for cases where orthopedic surgeons must fuse bones in a small child's hand or foot.

To solve these problems, custom-designed bone plates can be used. As proof of concept, two projects were carried out at North Carolina State University in a collaboration between the Industrial Engineering Department and the College of Veterinary Medicine. The first project looked at a custom-designed bone plate for distal femur fractures which are very common in both young children and animals. This is a type of fracture where the distal femur fracture is located close to the growth plate. The fracture must be fixated using either wires/pins or bone plates. As mentioned earlier, the manual shaping of bone plates can be a very time-consuming task that prolongs the surgical procedure and increases the risks for the patient. For this project, a CT scan was acquired of the patient, and the images were imported into the Mimics software package. The CT images were converted into a 3D model of the patient's fractured distal femur and were then exported in STL file format. Geomagic Studio from Raindrop Geomagic, United States, was used to convert the STL file into a NURBS model suitable for import into a CAD program. The CAD model of the distal femur was used as the basis for the custom-designed bone plate. The surgeon examined the CT scans and the 3D model in order to determine the optimal locations and angles for the screw holes. The engineering collaborators then designed a bone plate based on the recommendations of the surgeon. The bone plate was designed to conform to the shape of the bone without interfering with the surrounding soft tissue, ligaments and tendons. Figure 9.4 shows a CAD rendering of the custom bone plate, and Figure 9.5 shows an SLA model of the bone plate attached to SLA bone models following a mock surgery by the surgeon.

In a second project, a custom bone plate was designed to fixate the proximal tibia after a closing wedge osteotomy. In this case, both a custom bone plate and a custom drill and cutting guide were designed. A closing wedge osteotomy on the tibia is commonly used to correct the tibial plateau slope to prevent femoral-tibial dislocations. There are five common procedures for performing a tibial plateau leveling osteotomy (TPLO), and three of them require a bone plate to fixate and stabilize the bone. The cuts performed during a closing wedge osteotomy are normally done free-hand by the surgeon without any type of guide. It takes a very skilled surgeon to achieve accurate results with this procedure. After the cut is made, the bone plate must be shaped by hand to follow the contours of the bone. To reduce surgery time and to increase the accuracy of the cut and placement of the bone plate, the surgeon gave recommendations prior to the surgery regarding the optimal cutting location and the required realignment of the

Figure   9.4 CAD model of the custom-designed bone plate, showing the front and the backside

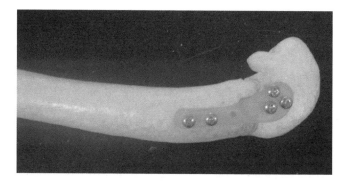

Figure   9.5 SLA model of the fractured distal femur and the custom-designed bone plate

tibial plateau. The surgeon further specified the overall size of the plate and the optimal screw locations. A CT scan of the patient's leg was acquired, and a 3D CAD model of the tibia was created using the procedure as described in the previous case. A custom cutting and drilling guide was designed that would provide the accurate alignment of the tibial plateau as well as the correct hole locations (Figure 9.6). A custom bone plate with a matching hole pattern was designed that would effectively fixate and stabilize the bone. The concept of a custom-designed cutting and drill guide based on patient-specific CT data is similar to SurgiGuide developed by Materialise, Belgium, and is covered in detail in Chapter 4.

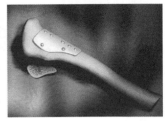

Figure 9.6 Rendered images of the custom cutting guide, the cut tibia and the custom bone plate

9.3 Electron Beam Melting Technology

After a custom implant is designed, the next step is to fabricate the device in an appropriate material. CNC machining from bar stock is a common fabrication method for custom implants. Since custom components have complex geometric shapes, a five-axis CNC machine is usually needed to do the fabrication. Generating five-axis CNC toolpaths is quite complicated, and fixturing complex shapes during multiple set ups requires considerable skill and experience. Once the toolpath is generated, the cutting operation itself can take a very long time owing to the amount of material that needs to be removed from the solid bar stock. Many orthopedic implant devices are made out of titanium or cobalt–chromium, both of which are very expensive materials. When a component such as a knee implant is machined from solid bar stock, as much as 80% of the bar stock is converted to metal chips. Between the lengthy machining times, the expensive five-axis CNC equipment and the large quantities of scrap materials, it is easy to see that CNC machining of custom implants from bar stock is not a promising approach for mass customization of implant components.

The remainder of this chapter describes a solid freeform fabrication (SFF) method that is being examined as a more cost-effective way to produce custom implant components. Even though all SFF technologies are able to fabricate the complex shapes of a custom-designed component, very few are capable of fabricating components directly out of biocompatible metals. The electron beam melting (EBM) process developed by Arcam (www.arcam.com) is one such process. EBM is a solid freeform fabrication process that builds fully dense metal parts in layerwise fashion. The process starts by spreading a thin layer of metal powder over a build platform. The layer thickness is typically 0.12 mm for titanium powder, although the thickness is user selectable. Next, an electron beam with up to 4.8 kW power is focused to a spot size of approximately 0.1 mm diameter on the surface of the powder bed, thus melting whatever metal powder it impinges upon. The melt pool produced by the beam is approximately 1.2 mm wide. The beam is scanned in the cross-sectional image of the layer being fabricated until that layer is completed. The platform is lowered, a new layer of powder is spread and the process is repeated one layer on top of the next until the part is complete.

The basic configuration of the EBM machine is shown in Figure 9.7. The electron beam gun is located above the build chamber. The electron beam originates at the tungsten filament and is accelerated to approximately half the speed of light by the anode. An electromagnetic coil focuses the beam to a spot on the powder bed, and the deflection coil moves the beam from one location to another on the powder bed. During processing, the build chamber is held under a vacuum of approximately 1×10^{-3} mbar and the gun is held under a vacuum of

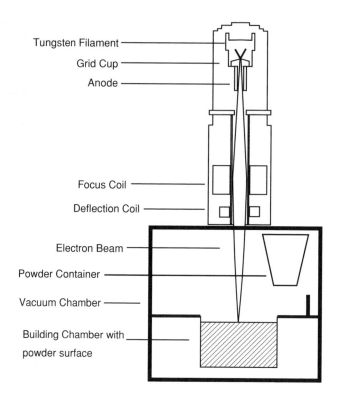

Figure   9.7 EBM machine configuration (courtesy of Arcam)

approximately 1×10^{-6} mbar. Although this is done to prevent scattering of the electron beam from collisions with air molecules, it has the added benefit of helping to purify the metal by preventing high-temperature oxidation and by removing vaporized impurities.

Each layer is processed in three distinct phases. The first phase of processing is the melting phase. Layers to be melted are subdivided into two regions – contours and squares. The meaning of contours is obvious. Users may specify that the beam scan multiple contours, which are offset by any desired amount. A typical setting would be two contours that are offset by 0.3 mm. Since the melt pool is approximately 1.2 mm in diameter, specifying a contour offset of 0.3 mm means that the melted contours will overlap one another by 0.9 mm.

For any given layer, the remaining area not taken up by melted contours is subdivided into a grid of squares. The user specifies how large these squares are, with square sizes typically ranging between 10 and 15 mm. In order evenly to distribute heat throughout the layer and thus minimize problems with distortion from shrinkage, the control software directs the electron beam to melt these squares in random order. The beam current and scan speed used during melting are very much dependent on the part geometry. Thin-walled delicate features are generally melted using lower beam currents in the 5–10 mA range, whereas large bulky areas may be melted much more quickly using higher beam currents in the 20–30 mA range.

The final phase of processing is known as post-heating. During post-heating, the electron beam rapidly scans the powder bed surrounding the area that has just been melted. This heats

Table 9.1 Ti6Al4V material properties

Hardness	HRC 30–35
Tensile strength	930 MPa
Yield strength	880 MPa
Elongation	10%

up the entire area surrounding the part, and the next layer of powder is then quickly spread on top of this heated area. A delay of several seconds prior to melting of the next layer allows the newly spread powder to heat up and lightly sinter to the previous layer. This post-heating procedure serves two functions. By lightly sintering the powder particles in place, the particles resist the tendency to repel one another and fly away when they are bombarded with a stream of negatively charged electrons. The second purpose of the post-heating sequence is to improve the quality of downfacing surfaces. When a downfacing surface is being melted on top of loose powder, the beam penetrates more deeply and can essentially blow small holes into the powder bed. Molten material above can penetrate into these holes to produce spikes or 'icicles' protruding from downfacing surfaces. When the powder beneath a downfacing surface is lightly sintered during post-heating, the beam does not penetrate as deeply and does not displace the powder. The icicles are eliminated, and downfacing surfaces are virtually indistinguishable from upfacing surfaces.

For the fabrication of biomedical implants, the material properties of the resulting powder are obviously important. At the present time, three powders are commercially available for use with the machine. They are pure titanium, titanium alloy (Ti6Al4V) and H13 steel [19]. Material properties for the Ti6Al4V material are listed in Table 9.1. It is important to note that these figures were obtained using standard testing procedures. Work remains to be done to determine whether or not the material properties for thin-walled implants are in agreement with material properties obtained with the more bulky tensile bars.

9.4 Direct Fabrication of Titanium Orthopedic Implants

9.4.1 EBM fabrication of custom knee implants

The introduction of electron beam melting technology and the development of the titanium material have resulted in a new fabrication method for custom-designed titanium implants and devices. The EBM machine is capable of building parts directly out of titanium to near net shape that require very little finish machining. After final machining, the parts can be polished to a mirror finish to provide excellent bearing surfaces for implants. As is the case with many other SFF systems, multiple components can be nested within the build chamber. Depending on the size of the implants being built, approximately 8–12 different custom titanium knee implants can be fabricated in the same build. Using currently recommended process parameters, it takes approximately 10–12 h to build a nested batch of knee implants. This equates to a production rate of 60–90 min per implant per machine. Assuming a machine produces two batches of implants per day for 5 days per week, one EBM machine is capable of producing over 4000 custom knee implants per year. The EBM fabrication and material costs add up to approximately

US$150–200 before light finish machining. Although this is not inexpensive, it is far more time and cost effective than machining custom implants from bar stock.

The EBM process was originally developed with the tool and die making industry as its primary target. Steel tools are typically large and bulky objects in comparison with relatively thin-walled knee implants. In order to test the feasibility of making thin-walled custom implants via the EBM process, a custom knee implant was first designed using procedures discussed earlier in this chapter. The femoral component was then fabricated in H13 steel on the EBM machine in order to assess whether or not the machine was capable of fabricating a thin-walled component. Figure 9.8 shows the implant that was produced after hand finishing and polishing.

Once it was established that the process was capable of producing thin-walled implants, a custom implant for a canine Labrador Retriever with arthritic knees was designed in collaboration with orthopedic surgeons in the College of Veterinary Medicine. The implant was then fabricated in titanium Ti6Al4V alloy. The hand-polished titanium implant is shown in Figure 9.9.

9.4.2   EBM fabrication of custom bone plates

Once the feasibility of fabricating thin-walled knee implants via the EBM process was established, a new project involving fabrication of custom bone plates was initiated. The custom bone plate was designed to fit a dog's tibia following a tibial plateau leveling osteotomy (TPLO). This is a surgical procedure in which a wedge of bone is removed from the tibia. The two halves of the tibia are then held together via a bone plate until the bone heals over the course of approximately 2 months.

For this project, the surgeon identified the location where the cuts would take place. A CAD model of the cut and reassembled bone was created, and a bone plate conforming to this new bone surface was designed. The surgeon also specified the number and location of screw holes.

Figure   9.8 Custom-designed femoral component fabricated in tool steel

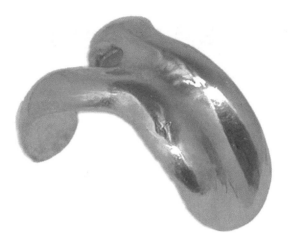

Figure 9.9 Custom-designed femoral component for a dog, fabricated in titanium

Following this design phase, the bone plate was built in the EBM machine standing on edge. Although this increases the total build time, it reduces the well-known stair stepping effect associated with nearly horizontal surfaces. Figure 9.10 shows the bone plate after the outer surface has been polished to prevent friction between the plate and the surrounding soft tissue. In order to test the fit of the custom bone plate, a model of the patient's bone was built in plaster on a Z-Corp machine. The model was infiltrated with cyanoacrylate to add strength. The surgeon then cut the bone model and removed a wedge of material as he would do in surgery. As shown in Figure 9.11, the separated halves of the bone were then attached by the bone plate. Visual inspection indicated that a very close fit was achieved without the need for hand-fitting by the surgeon.

9.4.3 Direct fabrication of bone ingrowth surfaces

The bone–implant interface on cementless implant components is typically modified to produce a porous surface that enables bone ingrowth. This mechanical interlocking between the bone and implant results in a more secure and stable prosthesis. The porous bone–implant interface

Figure 9.10 Custom-designed bone plate fabricated using EBM technology

Figure   9.11 Bone plate installed on a corrected tibia

on cementless implant components is typically produced by sintering small titanium beads to the surface. To produce enough porosity, the sintering process is repeated 3 times. There has been much research done on the properties of the porous surface. Different manufacturers use different-size titanium beads, and some companies have developed new processes to increase the porosity.

An alternative method for producing the porous bone ingrowth surface on custom implants using the EBM process is currently being investigated. With this approach, the power of the electron-beam is decreased and/or the scan speed is increased such the titanium powder does not have sufficient time or temperature to densify fully. The result is an interconnected network of pores in the metal component. Preliminary experiments indicate that a very wide range of pore sizes and shapes can be obtained by varying the intensity and duration of powder heating. The optical micrograph shown in Figure 9.12 shows a sample with interconnected pores in the 0.05–0.1 mm size range. Research is currently being conducted to compare the achievable pore morphologies with those that are obtained using the traditional bead sintering approach.

9.5   Summary and Conclusions

This chapter has described efforts to assess the feasibility of using the electron beam melting process to fabricate custom-designed biomedical implants. Although the EBM process was originally intended to produce large bulky tooling components, the feasibility of producing thin-walled implant components has been successfully demonstrated. The process has been

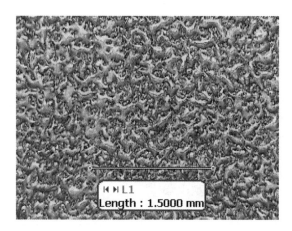

Figure   9.12 Porous network created via the EBM process

used to fabricate custom human and canine knee implants as well as a custom bone plate for a TPLO surgery. Both steel and titanium alloy materials have been used. Future development efforts will target both cobalt–chromium and stainless steel materials that are commonly used for orthopedic components.

Although the technological feasibility of fabricating custom biomedical implants has been established, an equally important consideration is cost. While it is clear that custom implants will cost more than mass-produced generic implants, the additional cost is not expected to be prohibitively large provided the custom implants are produced in batch mode on the machine. Although insurance companies typically focus on short-term immediate costs, it is also important to recognize the potential for custom implants to reduce long-term costs associated with revision surgeries.

The next stage of research with this process will involve moving towards human trials. This will first involve demonstrating that the material being used is sufficiently close in composition to the titanium alloys already approved for implant use for it not to have to undergo additional biocompatibility testing. Relevant material properties for thin-walled implants such as yield strength and stiffness will be tested. Particular attention will be paid to the effect, if any, of build orientation on these material properties. Next, the design process, the fabrication steps and the surgical procedures will be tested and refined on animal subjects in conjunction with colleagues in the College of Veterinary Medicine. Canine patients with severely arthritic knees are routinely seen at this hospital, and no practical treatment method currently exists for this condition in canines. Once the surgical procedure has been sufficiently tested and refined using canine patients, trials with human patients will commence.

References

Harrysson, O., Robertsson, O. and Nayfeh, J. (2004) Higher cumulative revision rate of knee arthroplasties in younger patients with osteoarthritis. *Clin. Orthop. and Related Res.*, **421**, 162–8.

Toksvig-Larsen, S. and Ryd, L. (1994) Surface characteristics following tibial preparation during total knee arthroplasty. J. Arthroplasty, **9**, 63–6.

Fadda, M., Marcacci, M., Toksvig-Larsen, S., Wang T. and Meneghello, R. (1998) Improving accuracy of bone resections using robotic tool holder and a high speed milling cutting tool. J. Med. Eng. and Technol., **22** (6), 280–4.

Pokrandt, P., Both, A. *et al.* (1999) Computer assisted surgery planning And robotics by orto MA-QUET. 4th International Workshop on Rapid Prototyping in Medicine and Computer-Assisted Surgery, Erlangen, Germany.

Keenan, J., Chakrabarty, G. and Newman, J. H.(2000) Treatment of supracondylar femoral fracture above total knee replacement by custom made hinged prosthesis. *The Knee*, **7**, 165–70.

Sathasivam, S., Walker, P. S., Pinder, I. M., Cannon, S. R. and Briggs, T. W. R. (1999) Custom constrained condylar total knees using CAD-CAM. *The Knee*, **6**, 49–53.

Joshi, A. B., Lee, J. and Christensen, C. (2002) Results for a custom acetabular component for acetabular deficiency. J. Arthroplasty, **17** (5), 643–8.

Werner, A., Lechniak, Z., Skalski, K. and Kedzior, K. (2000) Design and manufacturing of anatomical hip joint endoprostheses using CAD/CAM systems. J. Mater. Process. Technol., **107**, 181–6.

Viceconti, M., Testi, D., Gori, R., Zannoni, C., Cappello, A. and De Lollis, A. (2001) HIDE: a new hybride environment for the design of custom-made hip prosthesis. *Comput. Meth. and Programs in Biomed.*, **64**, 137–44.

Wolford, L. M., Pitta, M. C., Reiche-Fischel, O. and Franco, P. F. (2003) TMJ concepts/Techmedica custom-made TMJ total joint prosthesis: 5-year follow-up study. Int. J. Oral and Maxillofac. Surg., **32**, 268–74.

Harrysson, O. L. A. (2001) PhD dissertation, University of Central Florida, Orlando, FL.

Cormier, D. R. and Harrysson, O. L. A. (2003) Contoured machining of orthopedic surfaces for custom biomedical implants. IERC 2003, May 18–20, 2003, Portland OR.

Jajal, K. R. (2003) Custom design and manufacturing of canine knee implants. Master's thesis, North Carolina State University, Raleigh, NC.

Harrysson, O. L. A., Cormier, D. R. and Jajal, K. (2003) Custom design and manufacturing of canine knee implant. IERC 2003, May 18–20, 2003, Portland OR.

Harrysson, O. L. A., Cormier, D. R., Marcellin-Little, D. and Jajal, K. (2003) Direct fabrication of metal orthopedic implants using electron beam melting technology. Solid Freeform Fabrication Symposium 2003, August 4–6, 2003, Austin, TX.

van Steenberghe, D., Malevez, C., Van Cleynenbreugel, J., Serhal, C. B., Dhoore, E., Schutyser, F. Suetens, P. and Jacobs, R. (2003) Accuracy of drilling guides for transfer from three-dimensional CT-based planning to placement of zygoma implants in human cadavers. *Clin. Oral Implant Res.*, **14**, 131–6.

Eufinger, H. and Saylor, B. (2001) Computer-assisted prefabrication of individual craniofacial implants. *AORN J.*, November, **74** (5), 648–54.

Hieu, L. C., Bohez, E., Vander Sloten, J., Oris, P., Phien, H. N., Vatcharaporn, E. and Binh, P. H. (2002) Design and manufacturing of cranioplasty implants by 3-axis cnc milling. *Technol. and Health Care*, **10**, 413–23.

Cormier, D. R., Harrysson, O. L. A. and West, H. (2004) Characterization of high alloy steel produced via electron beam melting. *Rapid Prototyping J.*, **10** (1), 35–41.

10

Modelling, Analysis and Fabrication of Below-knee Prosthetic Sockets Using Rapid Prototyping

J. Y. H. Fuh, W. Feng and Y. S. Wong

As the shape, size and bone location of the stump of an amputated limp vary from one to another, the fabrication of a below-knee (BK) prosthetic socket has to be customized, which is currently a very time consuming and laborious process. This chapter presents a computer-facilitated approach to customizing BK prosthetic sockets, which first involves computer-aided modelling and analysis of the prosthetic socket design, and finally the fabrication of the socket based on the rapid prototyping technique.

Firstly, a finite element model of the BK prosthetic socket is developed to investigate the stress distribution during static loading. Subsequently, for experimental verification, an acrylic prosthetic part and a plastic prosthetic part are fabricated using the stereolithography and FDM processes respectively. The experimental results are compared with those obtained using finite element analysis (FEA).

A specially built RP system for the fabrication of prosthetic sockets, called the rapid socket manufacturing machine (RSMM), is also presented. The data capture software and techniques used in the system are also described.

Advanced Manufacturing Technology for Medical Applications Edited by I. Gibson
© 2006 John Wiley & Sons, Ltd.

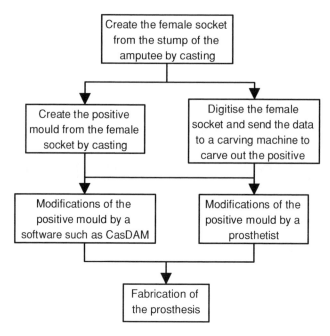

Figure   10.1 Process flow in making a prosthetic socket

10.1   Introduction

Fabrication of the prosthetic socket for below-knee (BK) amputees has always been a time consuming and complicated task. As the conditions of each patient's stump differ, each socket has to be customized. Thus, much expertise is also required on the part of the prosthetist. Because of its freeform shape, it is not an easy task to perform design analysis and automate the fabrication process of the socket. The process flow can be summarized as shown in Figure 10.1 (Otto Bock, 1996).

10.1.1   Process of making the below-knee artificial prosthesis

The process first involves soaking a bandage in water and plaster of Paris, and then wrapping the layer of bandage around the stump of the patient to obtain its shape. The bandage/plaster composition is then allowed to harden. The hardened socket is removed from the stump and used as a negative mould. Plaster of Paris is poured into the negative mould. Some time is again allowed for the plaster to harden before the negative mould is cut open to extract the positive mould within. Finally, the surface of the positive mould is smoothened. Alternatively, a special digitizer is used to digitize the inside of the negative mould. The digitized data are then sent to a CNC machine to carve out the positive mould.

10.1.1.1   Shaping of the positive mould

First of all, the sensitive regions on the stump, which cannot take too much load, must be identified, for example, regions by the sides of the knee and the distal end of the stump (shaded

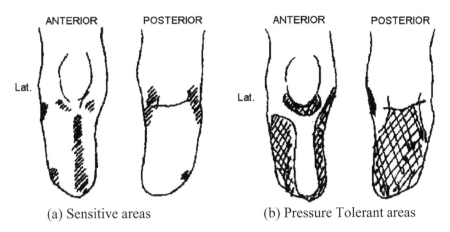

(a) Sensitive areas   (b) Pressure Tolerant areas

Figure   10.2 Sensitive and pressure-tolerant areas on the stump

areas in Figure 10.2a). At these locations, relief (allowance) has to be provided so as to prevent any stressful contact between these sensitive regions and the prosthetic socket. This can be achieved by applying additional plaster to these regions, so that the prosthetic socket produced will have recesses and not have a tight fit over the stump in these regions. Otherwise, this can also be left alone because, during the production of the prosthetic socket, a layer of fabric will be wrapped over the positive mould before the epoxy is poured over it. Hence, when the fabric is removed, allowance is introduced automatically.

In a region where it can take load, for example, just below the knee (shaded areas in Figure 10.2b), more of the material is removed so that the prosthesis will sit right on these areas which can serve as supports. The shaping of the positive mould is a process that requires expertise. The prosthetist will feel the bone structure of the patient and decide the amount of material to be removed in the pressure-tolerant areas. There are software packages that can be used to facilitate these operations. The software will take in the CAD data and perform the necessary modifications using expert design formulae (a kind of expert system). There is a list of expert prosthetists from which a user can choose to make the necessary modifications. Hence, the output can be quite different, depending on the chosen prosthetist, but the general guideline is based on the steps explained above.

10.1.1.2   Fabrication of the prosthesis

In the fabrication of a prosthetic socket, a layer of fabric is firstly wrapped over the positive mould. Subsequently, a sheet of material is placed over the fabric. Two kinds of material can be used in this case: acetate sheet or polypropylene sheet. Epoxy is then poured into the space between the fabric and the sheet of material. By using a vacuum system to suck out the air, the epoxy is drawn into the fabric. After allowing the epoxy to harden, this procedure is repeated another 4–6 times, depending on the size and weight of the patient. When the prosthetic socket is fabricated, the shaft and the artificial foot are attached onto it to produce the lower-limb prosthesis. Finally, the prosthesis is suitably finished to give it a more genuine look, (see Figure 10.3).

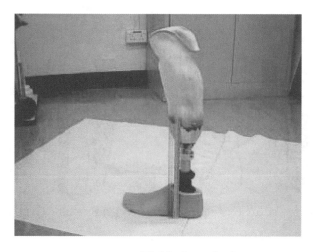

Figure   10.3 Finished prosthesis

10.1.2   Modelling, analysis and fabrication

With the FEA and RP techniques, the aforementioned tedious process can potentially be simplified and shortened. The long-term goal is to be able to develop an analytical tool that can be used to predict the stress distribution of a prosthetic socket design and fully automate the process of fabricating a prosthesis. A computer-aided method of fabrication by the stereolithography (SL) process has been proposed to automate and simplify the process so as to increase productivity (Sanders and Daly, 1993). Using this method, the prosthetic socket is synthesized via the photopolymerization of a suitable photosensitive monomer. The SL process relies on a scanning ultraviolet laser beam to harden successive thin layers of photopolymer, building each layer on top of the previous layer until a three-dimensional part has been formed. With the development of durable and strong materials suitable for photopolymerization, the stereolithography method of fabrication has a high chance of producing functional prosthetic parts and at the same time shortening the production lead time.

FEA is the process of breaking up a domain into a finite number of elements and solving the discretized domain. Division of the domain is accomplished by overlaying a suitable imaginary mesh. FE domains may consist of physical objects, such as mechanical parts. With FEA it is possible to predict, with a certain degree of accuracy, responses of the object when subjected to external actions. Computer models of the residual limb below the knee have been utilized to investigate the prosthetic socket interface stresses. Numerical analysis or FEA of the prosthesis offers several advantages over experimental measurements in the estimation of prosthesis interface stresses (Sanders and Daly, 1993; Silver-Thorn and Childress, 1996; Quesada and Skinner, 1991; Sanders, Daly and Burgess, 1993). For example, the stress or displacement distribution of the whole prosthesis can be examined. In addition, different prosthesis designs, characterized by material modifications, can also be easily investigated prior to actual fabrication of the prosthesis. This is especially applicable to the design stage.

Comparison of FEA results with experimental interface stress measurements during the stance phase of gait has often been conducted, and the shear and normal stresses have been measured at some sites on the stump, for example, the patellar tendon, antero-lateral proximal, antero-lateral distal, postero-proximal, postero-distal and antero-medial proximal (Sanders and Daly, 1993). Investigations have also been made of the effects of parameter variations on the interface stress distribution during static stance phase on a model based on geometric approximations. The results indicate that the residual limb/prosthetic socket interface stresses are affected by variations in both the prosthesis design and the residual limb geometry, which include the stiffness of the prosthetic liner and soft tissue, prosthetic rectification, the shape and bulk of soft tissue and residual limb length (Silver-Thorn and Childress, 1996). Small changes in stump length (2 cm) can cause relatively large pressure changes (16–18 %) (Quesada and Skinner, 1991). Shear stress is another factor to be taken into consideration (Sanders, Daly and Burgess, 1993). When applied simultaneously with normal stresses, they can cause injury within the skin. Hence, the prosthetic socket must be designed specifically so that it fits well without causing any pain to the patient through undesired shear or normal stress at certain sites on the stump.

10.2 Computer-Facilitated Approach

As the moulding process is very tedious, the aim is to automate it as much as possible. This can start with CT or MRI scan data of the stump and conversion of the dataset into a CAD model or STL file. A software called 'Mimics' is used to convert the CT or MRI files into contour or STL files, and some other CAD software can read in these contour files and produce a CAD model. One advantage of this method is that it will also be able to provide data about the internal structure, e.g. the bone or tissue, which is useful for FEA. Finally, the CAD model is processed and sent to the SLA machine for the fabrication of the prosthetic socket. Figure 10.4 below is a diagram showing both the traditional and the proposed process flows of producing the final prosthesis.

10.2.1   CAD modelling

A three-dimensional CAD model of a left patellar-tendon-bearing (PTB) below-knee (BK) prosthetic socket with a length of approximately 20 cm is constructed from a modified positive mould of a stump. The mould is digitized in a layer-by-layer manner into contour data using a coordinate measuring machine (CMM). Altogether there are 46 contours produced at varying height, with more contours taken at areas with complicated surfaces, eg. at the knee level. The contour dataset is then read in by a software called 'Surfacer' (ImageWare 1996). Using Surfacer, all the contour data are converted into closed curves. A surface is then created by lofting these curves. Figure 10.5a shows the surface model of the prosthetic socket.

An external surface is also created by lofting another set of curves created by offsetting the original curves in the X–Y direction by a constant distance of 3 mm. By doing so, these two surfaces serve to enclose a thin shell with a thickness of 3 mm. Although the FEM package is able to analyse the model as a shell element, further modifications to the thickness at a particular location is not permissible. However, with the two surfaces, localized modifications of the thickness can be easily made to the external surface without altering the internal surface.

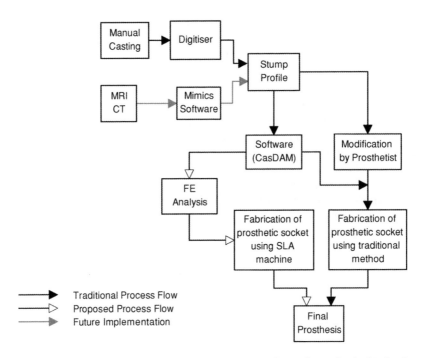

Figure **10.4** Traditional and computer-aided process flow of prosthesis fabrication

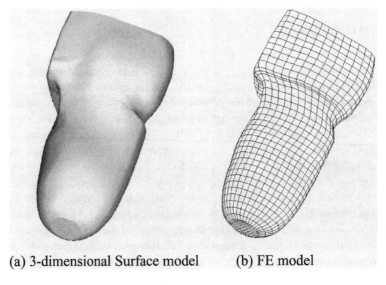

(a) 3-dimensional Surface model   (b) FE model

Figure **10.5** Surface and FE model of the BK prosthetic socket

10.2.2   Finite element analysis (FEA)

10.2.2.1   Geometries

An FE meshed model (Figure 10.5b), with 1396 nodes and 1377 elements, was generated using 'Patran', an FE software, with a size of about 1 cm by 1 cm for a general element. Four-node quadrilateral shell elements were used. Evaluation runs were conducted to ensure that elemental edge angles and elemental face warpage were within a 45° angle, except at locations where the results were not important and did not significantly affect the overall stress distribution.

10.2.2.2   Boundary conditions

In an actual prosthesis, a wooden base is fixed to the distal end of the prosthetic socket. The other side of the base is flat and a shaft, together with the artificial foot, is connected to it. As the shaft is of considerable stiffness compared with the socket, we assumed a fixed boundary condition to the entire distal portion of the socket up to a distance of about 3 cm from the tip of the distal end. The rest of the model was assumed to have complete free movements.

10.2.2.3   Loading conditions

Loading was based on an approximate height of 170 cm and weight of 70 kg for the aforementioned adult prosthesis (Maquet, 1983). Figures 10.6 and 10.7 below show the biomechanics

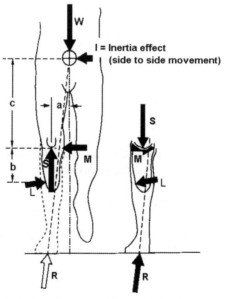

(a) Forces on the Amputee   (b) Forces on the prosthesis

Figure   10.6 Mediolateral force diagram of a PTB BK prosthesis

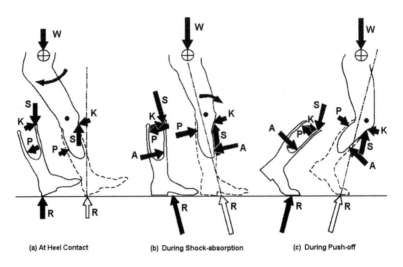

(a) At Heel Contact   (b) During Shock-absorption   (c) During Push-off

Figure 10.7 Anteroposterior force diagram of a PTB BK prosthesis

of a patellar-tendon-bearing below-knee prosthesis. Figure 10.6 illustrates the forces acting on the amputee and prosthesis in a mediolateral view, and Figure 10.7 in an anteroposterior view.

In Figure 10.6, the reaction force L is obtained by calculation from the following static equilibrium equations:

$$\sum M_k = 0$$

and thus

$$L_b + I_c = W$$

$$L = \frac{W_a - I_c}{b}$$

$$M = L - I$$

Where

    S, L and M are forces acting between the socket and the stump;
    W is the gravitational force acting on the body (or weight);
    I is the lateral inertia force;
    R is the reaction from the ground to the prosthesis.

For the application of loading conditions on the FE model, particular attention was paid to ensuring that areas that were not pressure tolerant were free from applied load.

10.2.2.4   Analysis

For the analysis, other FEM software called 'Abaqus' (Abaqus, 1995) was employed. It read in an input file to 'Abaqus' produced from 'Patran' to perform the FEA. The displacement or stress distribution can be visualized using 'Abaqus Post'. If the analysis results are satisfactory, fabrication can begin. Otherwise, modifications are made to the area of high-stress concentration and the FE analysis is run again.

10.3   Experiments

After the iterative process indicates satisfactory stress distribution from the analysed results of the model, an STL file of the model is output to an SLA machine to fabricate a functional prosthetic socket. The fabricated prosthetic socket used to verify the FEA had a shell thickness of 4 mm. Compression tests were conducted on it using an Instron servohydraulic testing machine. The strain data were then compared with the FEA results.

Load testing of the prosthetic socket required the amputee's remaining stump to be fitted into the prosthetic socket. However, this could not be done since the amputee was not available. Thus, a substitute stump was made to simulate loading conditions by solidifying a silicone-based resin in the socket. The internal bone structure was simulated by placing a corresponding part of the knee bone in the socket before the resin was poured in. Subsequently, the resin was allowed to solidify with the bone structure intact. Some parts of the 'stump' were shaved off so as to shape it as close to the shape of the real stump as possible. The silicone resin used to make this substitute stump was soft so as to simulate flesh. This 'stump' was then fitted into the socket. The distal part of the socket was fixed onto a wooden base (20 cm by 15 cm) since the socket was to be attached to the shaft and thus assumed to be fixed.

The material characteristics of the UV-curable acrylic-based resin used are assumed to be linear, homogeneous and isotropic. Table 10.1 below shows the properties of the material. The socket together with the 'stump' was subsequently placed in the Instron servohydraulic testing machine for load testing. The protruding bone was attached to the machine where the force was applied so as to simulate real-life conditions. Nine strain gauges were placed on the external surface of the socket where the strain was expected to be large. The strain gauges were of type GFRA-3-70, of 3 mm length. The positions of the strain gauges are shown in the diagram. The strain data were recorded at every 5 N force interval from 200 N up to 700 N.

The process flow between the creation of surfaces of the prosthetic socket to FE analysis of the model and finally to fabrication of the actual part is shown in Figure 10.8.

Table 10.1   Physical properties of the UV-curable resin used for the BK prosthetic socket

Properties	Values
Gel content	>98%
Elastic modulus in tension	1200 MPa
Tensile strength	39 MPa
Breaking extension	27%
Elastic modulus in bending	1360 MPa
Flexural strength	46 MPa
Hardness	HRC 92
Impact strength (with notch)	147 J/m
Impact strength (without notch)	461 J/m
Thermal deformation temperature	43°C

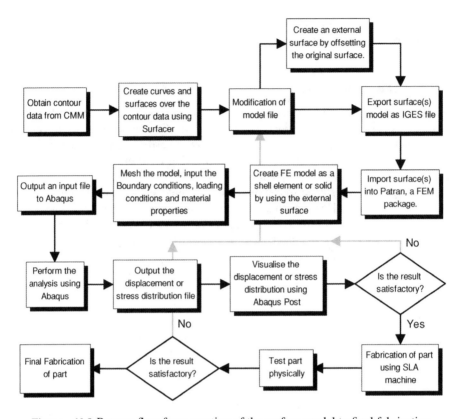

Figure   10.8 Process flow from creation of the surface model to final fabrication

10.4   Results and Discussions

Three FEA runs were made for the BK prosthetic socket with 3, 4 and 5 mm thick shell elements. The visual results on stress and strain produced using Abaqus Post are shown in Figures 10.9 and 10.10 respectively. Areas with the highest tensile stress are found at the lateral proximal, medial proximal and postero-distal sites, whereas the highest compressive stress is found at the antero-distal site.

Table 10.3 compares the 3 mm prosthetic socket and the material property of the post-cured resin. It can be seen that the maximum tensile or compressive stress is smaller than the flexural strength of the material by a factor of about 4–6, which is quite a safe margin to accommodate for any discrepancies in adding the loading.

For verification purposes, nine sites on the stump with predicted high stress and strain were chosen and strain gauges attached to them. The strain and stress data were then obtained and computed. The strain and stress at the point where the load cell reached 700 N are presented in Table 10.4. Figure 10.11 shows the positions of the strain gauges mounted for the experiment.

From the comparison it can be seen that there is a slight disparity between the FEA results and the experimentally obtained data, with the maximum difference in strain being 4.598×10^{-3} and the maximum difference in stress being 6.897 MPa. At site 3 (with reference to

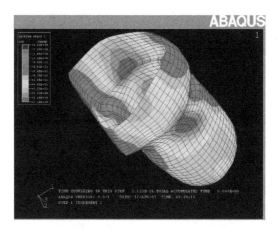

Figure   10.9 Stress distribution on a prosthetic socket

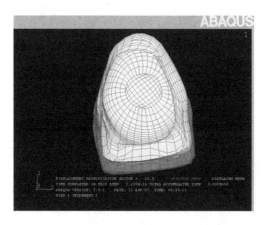

Figure   10.10 Displacement analysis

Figure 10.11) there is much less strain than expected. In fact, it has only a relatively small strain of 4.8×10^{-5}. Site 3 is a region where the stump is in constant contact with inner surface of the socket; thus, the strain should be preferably as small as possible to minimize abrasion to the skin around that area. In this respect the experimental tests showed better results than the FEA. At site 4 there is much more tensile stress than expected. This part of the socket is designed such that its surface is slightly protruding outwards, so as not to be in contact with the pressure-sensitive areas around the region. Thus, excessive tensile strain on this side of the socket should not have too much of an effect on the comfort of the patient. At Site 5 there is much tensile stress than expected (as FEA results predicted compressive stress and strain). Site 5 coincides with the area of the stump that is pressure-tolerant and thus may be able to withstand the high stress. However, further modifications may be needed. In summary, it must be noted that a simulated stump has been used instead of a real one. There is therefore room for experimental errors concerning the values obtained experimentally.

Table 10.2 Maximum and minimum stress, strain and displacement of the prosthetic socket with a shell thickness of 3 and 5 mm

	3 mm model			5 mm model		
Stress[a] (MPa)	S_1	S_2	S_{12}	S_1	S_2	S_{12}
Maximum	6.619	5.024	2.866	3.092	2.993	1.548
Minimum	-9.179	-8.938	-2.984	-5.362	-4.977	-1.479
Strain[b]	E_1	E_2	E_{12}	E_1	E_2	E_{12}
Maximum	4.034×10^{-3}	3.451×10^{-3}	7.022×10^{-3}	2.0284×10^{-3}	1.979×10^{-3}	3.792×10^{-3}
Minimum	-6.113×10^{-3}	-5.840×10^{-3}	-7.311×10^{-3}	-3.520×10^{-3}	-3.162×10^{-3}	-3.624×10^{-3}
Displacement[c] (mm)	U_1	U_2	U_{12}	U_1	U_2	U_{12}
Maximum	2.592	1.205	0.5952	1.183	0.2979	0.3053
Minimum	-5.333×10^{-2}	-1.459	-0.4815	-0.000	-0.4119	-0.2523
Rotation[d] (deg)	UR_1	UR_2	UR_{12}	UR_1	UR_2	UR_{12}
Maximum	1.4966×10^{-2}	1.5398×10^{-2}	2.7640×10^{-2}	7.2963×10^{-3}	4.8853×10^{-3}	7.4836×10^{-3}
Minimum	-1.0127×10^{-2}	-2.2113×10^{-2}	-3.5885×10^{-2}	-5.6582×10^{-3}	-8.6347×10^{-3}	-9.7539×10^{-3}

[a] S_{ij} = all-stress component. [b] E_{ij} = all-strain component.
[c] U_{ij} = all-displacement component. [d] UR_{ij} = all-rotation component.

Table 10.3 Comparison between the maximum stress and strain on the 3 mm prosthetic model and the material property of the post-cured resin

	3 mm thick model (a)	Post-cured resin (b)	Factor smaller by (b/a)
Stress (MPa)			
tensile	6.619	39	5.892
compressive	9.179		4.249
Strain (mm)			
tensile	7.022×10^{-3}	27	38.45
compressive	7.311×10^{-3}		36.93

Table 10.4 Experimental vs. simulation results

	Strain and stress at the different sites with load of 700N (70kgf)								
	Site 1	Site 2	Site 3	Site 4	Site 5	Site 6	Site 7	Site 8	Site 9
	Experimental results								
Strain $\times 10^{-6}$	1293	5055	48	2669	1378	552	143	13	44
Stress (MPa)	1.9395	7.5825	0.072	4.0035	2.067	0.828	0.2145	0.0195	0.066
	FE results								
Strain $\times 10^{-6}$	1359.9	1105.0	−374.19	−388.60	−417.98	695.91	546.66	−101.56	573.91
Stress (MPa)	2.142	1.447	−1.054	−0.9992	0.2849	0.2762	0.8701	−0.3269	0.5846

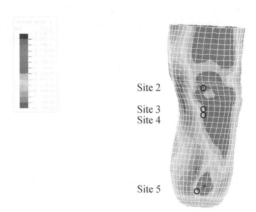

Site 2
Site 3
Site 4

Site 5

Figure   10.11 Strain distribution on the prosthetic socket (front view)

10.5   Rapid Socket Manufacturing Machine (RSMM)

As mentioned earlier, the traditional prosthetic socket fabrication process is tedious and long, besides being labour intensive, usually taking 2–3 days. On the other hand, the ability to create freeform surfaces and hidden features makes rapid prototyping (RP) an ideal technology for automated socket fabrication. However, to use SLA, FDM or any other commercial RP systems, there need to be practical means of customizing such a system for prosthetic socket fabrication, with process optimization to reduce significantly the manufacturing time and cost. As this may be difficult to accomplish at the present moment, a customized RP system has been specifically designed and developed for socket fabrication.

After investigating several different RP processes, the FDM process was chosen for the specially designed system because of its minimum post-processing requirement and the superior mechanical properties of the building material (P301 polyamide) that can be used.

10.5.1   RSMM design considerations

The requirements of the RSMM are different from the commercial RP system. Most of the building materials used by current RP systems are expensive and may not be suitable for the prosthetic sockets. Therefore, the building material for RSMM must be inexpensive and must meet the strength requirement of definitive sockets. Commercial RP systems usually have precision and accuracy that are beyond the requirement for socket fabrication. A diametrical accuracy of ± 1.0 mm is normally sufficient for prosthetic sockets. Therefore, the hardware and software design of the RSMM should make use of the lower accuracy requirement to reduce system cost and building time.

10.5.1.1   File format

STL is the most widely used format in rapid prototyping (Georges and Chuck, 1996). The STL model is formed by tessellation of the original model. It is excellent for describing models with planar surfaces, but in medical applications the limitations of STL models become obvious (Dolenc and Mäkelä, 1996). The highly complex shape of a prosthetic socket usually results in a large file size when tessellated. This in turn results in a long data verification and processing time, as in the making of the FDM socket.

As the STL format is apparently not a good format for representing prosthetic sockets, the RSMM has discarded the format. At this moment, the system accepts the following data formats:

- IGES (by Capod Systems/Digibotic);
- Point cloud (by Digibotic).

Additional data filters can be developed if required.

10.5.1.2   Nozzle

The nozzle used by the 3D Modeler FDM machine (Stratasys Inc., USA) has a relatively small diameter. Such configuration enables the machine to build small and delicate parts with good dimensional accuracy. However, the small nozzle diameter has its trade-off in terms of building speed. This is particularly apparent when building parts that are wider than the process road width.

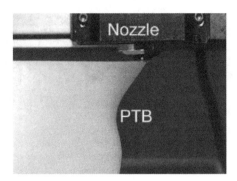

Figure **10.12** Building a slanted surface without external support

The road width of the FDM process ranges from 0.254 to 2.54 mm, depending on the diameter of the nozzle, the input speed of the material and the extrusion head pressure. In the case of making prosthetic sockets, the required socket wall thickness is usually thicker than the maximum road width. As it is impossible to achieve the required wall thickness in one pass, additional steps are needed to form a 'thicker' wall.

However, if the road width is designed to be the same as the required socket wall thickness, the fabrication time can be significantly reduced. To achieve this, the nozzle diameter has to be increased accordingly. This modification is justifiable for this application as building speed is the priority, and extreme precision in the $x-y$ plane is normally not required. The nozzle diameter used in the RSMM is 3 mm, with a resultant road width or socket wall thickness of 4 mm.

A wider road width also increases the processability to build slanted surfaces without external supports, as illustrated in Figure 10.12 for the patellar-tendon-bearing (PTB) region of the socket. Generating and building external supports can be extremely tedious and time consuming.

With a wider road width, the contact area between layers is also increased. This gives a better interlayer bonding, which is an important factor when socket strength is concerned. The interlayer bonding is further enhanced with the nozzle pressing the current layer against the previous layer as shown in Figure 10.13.

10.5.1.3 System accuracy

Besides manufacturing speed, socket accuracy is also an important issue. However, the accuracy requirement is not as high as for a commercial RP system. A diametrical accuracy of ± 1.0 mm is normally sufficient for prosthetic sockets. Therefore, the hardware and software design of the RSMM has made use of the lower accuracy requirement to reduce system cost and building time, but can meet the targeted system accuracy of ± 1.0 mm.

10.5.2   Overview of the RSMM

RSMM is similar to FDM, which dispenses semi-molten material onto the machine table, layer by layer, to form 3D objects. Polypropylene (PP) is selected to be the building material

Figure   10.13 Nozzle pressing the current layer against the previous layer

for the RSMM because of its rigidity, strength and resistance to fatigue. Most importantly, it is inexpensive and has been used by prosthetists for making above-knee and below-knee prosthetic sockets. There are two main components in the RSMM, namely the robotic system and the dispenser. A schematic and a photograph of the RSMM are shown in Figure 10.14.

A schematic of the dispenser is shown in Figure 10.15 Polypropylene filament (Ø4 mm) is fed into the heating barrel, The filament moves along the barrel and is melted near the exit where the heating element is mounted. The in-coming filament acts as a piston to push the molten polypropylene out of the nozzle.

The molten polypropylene forms a continuous strand as it leaves the nozzle. The strand is dispensed onto the machine table according to the cross-sectional contour of the stump. The second layer is laid in a similar manner on top of the first. The process continues until the whole socket is built (see Figure 10.16).

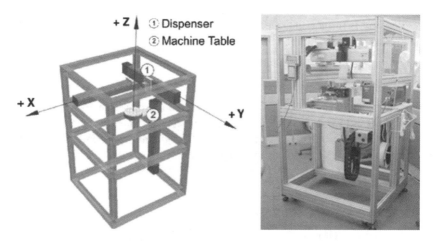

Figure   10.14 Overview of the RSMM

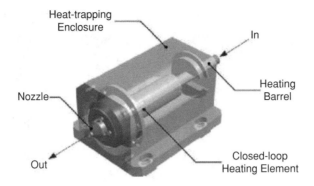

Figure   10.15 Schematic of the dispenser

Figure   10.16 Various stages in the manufacturing of a trans-tibial prosthetic socket

10.5.3   Clinical test

Clinical and biomechanical studies were conducted to evaluate the comfort and fit of the RSMM socket during gait. A prosthetic socket for a volunteer trans-tibial amputee subject was manufactured using the RSMM (see Figure 10.17). The time taken to build the socket was 3.5 h, and the complete prosthetic socket weighed approximately 1.6 kg, which was about 0.3 kg heavier than the traditional prosthesis. The comparison of the average temporal distance data of the stump wearing the RSMM and traditional socket is summarized in Table 10.5. Preliminary investigation of the RSMM socket showed that its functional characteristics were very similar to those of a traditional socket.

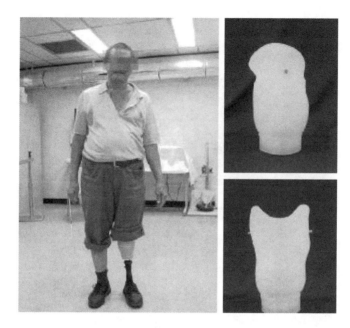

Figure   **10.17** A trans-tibial amputee with the RSMM socket

Table 10.5   Experimental vs. simulation results

	Traditional	RSMM
Cadence (steps/min)	91	86
Walking speed (m/s)	0.85	0.81
Stride time (s)	1.32	1.39
Step time (s)	0.64	0.68
Single support (s)	0.34	0.38
Double support (s)	0.54	0.54
Stride length (m)	1.13	1.11
Step length (m)	0.61	0.59
Stance (%)	66.67	65.83

10.5.4   Future work

The RSMM is a computer-aided rapid socket fabrication system that utilizes RP technology. The system reduces the socket fabrication time from days to less than 4 h. Prosthetic sockets are commonly reported to fail as a result of local bending and buckling at the distal region where the pylon is connected (Wevers and Durance, 1987). Therefore, to ensure that the RSMM socket is safe for use, the prosthetic socket is currently undergoing the ISO 10328 principal structure tests. More clinical and biomechanical studies will also be conducted in the future to evaluate the comfort and fit of the RSMM socket during gait.

10.6 Conclusions

An approach to improving BK prosthetic socket design based on FE analysis has been reported. A prosthetic socket model was fabricated using the RP process to shorten the fabrication time. It has shown potential for medical application while adopting the RP technique. Building the prosthetic socket using fibre-reinforced resin to increase its strength and toughness will enable it to be smaller in cross-section, and hence lighter.

The current software does not take into consideration the location of the bone remaining in the stump. Attempts will be made to convert the MRI or CT scan data of the stump into IGES files using 'Mimics'. With CAD models obtained from the scan data, the location of the bone can be considered in the design of the prosthetic socket. A more accurate analysis can then be performed to give a better modification of the prosthetic socket.

A customized RP system, the RSMM, developed to fabricate prosthetic sockets automatically, is also reported as a case study. The system reduces the socket fabrication time from days to less than 4 h.

Acknowledgements

Special thanks to Dr James Goh (Department of Orthopaedic Surgery, National University of Singapore) for providing some technologies and materials related to the RSMM system, and to the Advanced Manufacturing Lab in the National University of Singapore for providing the space and help to set up and maintain the RSMM system.

References

Abaqus (1995) *ABAQUS/Standard User's Manual*, Version 5.5, ABAQUS, Inc., Rhode Island.

Dolenc, A. and Mäkelä, I. (1996) Rapid prototyping from a computer scientist's point-of-view. *Rapid Prototyping J.*, **2** (2), 18–25.

Georges, M. F. and Chuck, K. (1996) Accuracy issues in CAD to RP translations. *Rapid Prototyping J.*, **2** (2), 4–17.

Imageware (1996) *SURFACER Training Guide, Version 5.0*, Imageware, Inc., San Diego.

Maquet, P. G. J. (1983) *Biomechanics of the Knee*, 2nd edn, Springer-Verlag.

Otto Bock (1996) *Otto Bock modular below knee prostheses*. Research and Development Department of Otto Bock Orthopadische Industrie GmbH Duderstadt, West Germany.

Quesada, P. and Skinner, H. B. (1991) Analysis of a below-knee patellar tendon-bearing prosthesis: a finite element study. *J. Rehabil. Res. and Dev.*, **28** (3), 1–12.

Sanders, J. E. and Daly, C. H. (1993) Normal and shear stresses on a residual limb in a prosthetic socket during ambulation: Comparison of finite element results with experimental measurements. *J. Rehabil. Res. and Dev.*, **30** (2), 191–204.

Sanders, J. E., Daly, C. H. and Burgess, E. M. (1993) Interface shear stress during ambulation with below-knee prosthetic limb. *J. Rehabil. Res. and Dev.*, **30** (2), 191–204.

Silver-Thorn, M. B. and Childress, D. S. (1996) Parametric analysis using the finite element method to investigate prosthetic interface stresses for persons with trans-tibial amputation. *J. Rehabil. Res. and Dev.*, **33** (3), 227–38.

Wevers, H. W. and Durance, J. P. (1987) Dynamic testing of below-knee prosthesis: assembly and components. *Prosthetics and Orthotics Int.*, **11**, 117–23.

Bibliography

Goh, J. C., Lee, P. V. and Ng, P. (2002) Structural integrity of polypropylene prosthetic sockets manu-factured using the polymer deposition technique. *Proc. Instn Mech. Engrs, Part H: J. Engineering in Medicine*, **216** (6), 359–68.

Lee, P. V. S., Tan, K. C., Tam, K. F. and Lye, S. L.(1998) Biomechanical evaluation of prosthetic sockets fabricated using fused depository method. The First National Symposium of *Prosthetics and Orthotics*, 1998, Singapore, pp. 41–42.

Ng, P., Lee, V. S. P. and Goh, J. C. H. (2002) Prosthetic sockets fabrication using rapid prototyping technology. *Rapid Prototyping J.*, **8** (1), 53–9.

Tan, K. C., Lee, P. V. S., Tam, K. F. and Lye, S. L. (1998) Automation of prosthetic socket design and fab-rication using computer-aided-design/computer-aided-engineering and rapid prototyping techniques. The First National Symposium of Prosthetics and Orthotics, 1998, Singapore, pp. 19–22.

11

Future Development of Medical Applications for Advanced Manufacturing Technology

Ian Gibson

11.1 Introduction

This book is a collection of works from renowned experts in the field of manufacturing technology and medicine. Each chapter has provided useful applications, case studies and insight into how this technology looks today and how it may evolve in the future. The purpose of this chapter is to re-address some of the more critical points. They may already have been raised explicitly or referred to in the previous chapters, and I make no excuse for repeating some of these. They may even conflict with some of the comments made by individual authors of chapters in this book, and again I offer no apology. Opinions may also differ from those held by readers of this book. It is not important who is right and who is wrong, but that the debate has started; that the issues have been raised and discussed and taken seriously. It is important that this field develops with improvements in technology that can subsequently improve patient treatment and quality of life.

Many of the readers of this book will be researchers who are looking for new ideas and new ways to develop their research. They may have (or indeed may be themselves) students who are looking for new topics to study or perhaps looking for a definition of the state of the art. This book will hopefully give some guidance on how to identify the important research areas and how to proceed. It is important to read the appropriate chapters of this book, perhaps followed by this chapter.

Advanced Manufacturing Technology for Medical Applications   Edited by I. Gibson
© 2006 John Wiley & Sons, Ltd.

I have divided the following sections according to the issues as I see them. I hope you find them helpful.

11.2   Scanning Technology

From a technological point of view this book starts with patient scanning systems. A number of systems have been presented, both conventional and unconventional. The most widely used systems for use with 3D modelling are CT and MRI. However, there are a number of applications that involve external scanning technology as well using laser and other forms of optical scanning equipment. Although touch-probe technology is widely used in industry, it is highly unlikely to be used for medical applications that involve patient scanning. Also, since anatomic data are highly freeform in nature, involving large point sets for accurate description of the surfaces and solids, all scanning systems must be highly automated.

However, such systems are currently large and bulky. There would be great benefit in making such technology portable, even to the extent of being able to mount the systems on a truck. For MRI, the magnet sizes are coming down for equivalent tesla ratings, but current systems are probably limited to military applications. Of course, most systems are designed to accommodate a full body. Most applications that require CT/MRI are head, back and pelvic regions, so the aperture dimension is always going to be quite large and the development of small machines that can only include arms and legs is somewhat limiting.

CT is still the more popular when compared with MRI. This is because it is based on a technology that has been with us for many years – X-ray technology. We all know that X-rays are dangerous to the human body, and CT requires interpolation of multiple X-ray images. Fortunately, the sensitivity of sensors has improved considerably in recent years, as well as the hardware and software used to process the raw data. However, there is always going to be a limit on the slice spacing of CT scanners that still results in a safe dosage. Currently, 3 mm is a normal lower limit. Where the application requires it, the more advanced CT scanners can be used to provide 1 mm slice separations. Even with this, RP machines are much more accurate. Therefore, the quality of the data is still currently limited by the scanning technology in this case. However, industrial CT and micro-CT (for non-living biosamples) both show the capability of this technology to work in the submillimetre and even micron scale.

MRI technology is most commonly used for soft-tissue analysis. CT is able to distinguish the tissue density, but, since MRI can be tuned to detect different molecules, the tissue structure can also be detected. This provides much clearer images of the patient data. Functional MRI provides the added benefits of monitoring the patient condition over time. However, there is a less distinct link between MRI technology and advanced manufacturing technology. This is perhaps because most AMT applications currently involve bony rather than soft tissue. Also, most applications assume homogeneous material properties. MRI is better suited to diagnostics and visualization. The main advantage is the ability to distinguish multiple materials and represent them in a virtual environment. RP technology is starting to catch up, with colour RP and also the ability to create heterogeneous structures. This is definitely a worthwhile research area to pursue.

One distinct advantage of external scanning technology is its portability. There are many laser scanners available that are low cost, accurate and easy to use. Where heads are being scanned, there are other methods that can be used to avoid endangering the eyes. Of course, the limitation is that there is no way to determine the internal structure of a sample, but there are numerous examples where the external cosmetic appearance of a patient may be required. A recent example is the interest forensic pathologists have in using portable laser scanners to capture 3D images of a victim at the scene of the crime.

Ultrasound imaging results in very low quality output compared with other systems. However, it is a very safe method that can perhaps find some applications. Noise removal procedures can be applied when sampling data over a long time period. It is also a relatively portable technology that can be used along with RP technology to visualize data that might be difficult to obtain in other ways.

11.3   RP Technology

The RP technology used to create medical models has proved invaluable in numerous applications described in this book. RP technology was initially developed to solve problems in manufactured product development. Certainly, the RP machine manufacturers recognize the value and benefit and they have gone some way to developing materials and process enhancements to suit the medical industry as well, but it is unlikely to become a priority. The following are some issues that distinguish medical applications from manufacturing:

- *Speed.* RP models can often take a day or even longer to create. The data preparation can in fact take even longer. This means medical models can only be included in surgical procedures that require long-term planning, and cannot be used, for example, as aids for emergency operations.
- *Cost.* Using RP models to solve manufacturing problems can help save millions of dollars for high-volume production. In comparison, medical models optimize the surgeon's planning time and improve quality and efficiency. These issues are more difficult to quantify in terms of cost, but it is clear that only the more complex cases can justify the expense of the models.
- *Accuracy.* Many RP processes are being improved to create more accurate components. However, medical applications currently do not require such high accuracy because the data from the 3D imaging systems are considerably less accurate than the RP machines into which they feed.
- *Materials.* Only a few RP materials are classified as safe for transport into the operating theatre, and none is currently capable of being placed inside the body. This limits the range of applications for models.
- *Ease of use.* RP machines generally require a degree of technical expertise in order to achieve good-quality models. This is particularly true of the larger, more complex and more versatile machines, which are also not particularly well suited to medical laboratory environments. Coupled with the software skills required for data preparation, this implies a significant training investment for any medical establishment wishing to use the technology.

For existing applications in medicine, to make the RP processes more effective, vendors must concentrate on developing machines that are inexpensive to run but with good material properties. Models need to withstand a degree of physical abuse including handling, cutting, drilling, etc. There are many machines currently available that provide sufficient strength. However, some parts that are strong enough may have a surface roughness that prevents the use of very small screws for practising fixation surgery. Most RP machines are accurate enough because the model data are generally less accurate than the RP process. They are all fast enough because applications requiring these processes are planned and scheduled over periods of weeks or months. Thus, the major pitfall for existing applications is the cost.

Of course, this does not describe the situation for developing new medical applications for RP. Emerging applications would benefit from:

- direct manufacture of biocompatible materials;
- heterogeneous manufacture, including multiple materials;
- colour, and other indicative features;
- integration with electronic and electromechanical devices;
- increased accuracy, to match the inevitable improvements in the scanning technology.

Of course, there is much emphasis from the computer scientists on the development of virtual reality for medical applications. VR certainly has a significant role to play in medical applications. There is a great deal of debate concerning which methods are best to use, and often it will result in personal preferences. However, it is clear that most doctors and surgeons are very tactile in the way they examine, diagnose and interact with patients and their ailments. Physical models made using RP are closer to the real-life situation in many respects. Spatial interpretation, incorporating tactile, haptic and visual feedback in a truly intuitive manner, is something that is very difficult to produce in a virtual environment. Eventually, this may be possible to achieve. However, the low-tech approach of using RP models is already with us.

11.4   Direct Manufacture

For many researchers into RP, the Holy Grail is rapid manufacture (RM). This is the process of taking the original CAD model and creating not only prototypes but also actual functioning parts. We live in a world where there is ever increasing demand for faster development of products, often with complex, unique or custom features. RP has the capability to fabricate parts directly from the original CAD data, without significant need to set up the machines with respect to the geometry of the part. This positions RP as an ideal technology for exploiting this demand. The layer approach can be applied to create components ready for use.

In a very limited sense, RM is already with us. An example has in fact been shown in this book for creating the stump connectors for artificial legs. However, at present, good examples are few and far between. RM can be an option for high-cost, customized, complex-geometry plastic parts. The stump connector example falls into this category and with further development could be considered a viable and cost-effective solution.

For most researchers, however, the main objective is to produce parts with good material properties and performance characteristics that would be difficult to obtain using more conventional manufacturing techniques. To many this means metals, with the added advantage of functional gradients and/or multiple material properties.

The capability to make good-quality metal components directly for use has yet to be properly achieved. With medical examples there is the added difficulty of creating parts in specific metals that have good functional properties and biocompatibility. One could argue that the use of RP to manufacture moulds and patterns for manufacturing parts using casting techniques achieves at least some of these objectives. The choice of metals could include those used for medical applications. However, creating these patterns and casting them successfully requires significant skills and resources, and it would be preferable to use a direct layer-based manufacturing process to make the required metal parts. It goes without saying that controlling a metal manufacturing process is considerably more difficult compared with the polymer systems commonly used in industry today. Most of the metal systems currently under development use a powder delivery system, which results in components that are relatively rough on the surface. While artificial joints might not require high precision in terms of the overall geometry, the roughness of any articulating surfaces must be extremely low. Post-processing of any component is likely to be high and may include surface treatments using chemical and electrochemical as well as mechanical methods. Researchers into RP and RM processes are aiming to develop them to include multiple materials and electronic components and features. Should this be achieved, then this would make a significant contribution to the creation of medical devices with sensors and intelligent features matched to an individual patient's body and needs. Other researchers into micron and submicron scale RP may also find many medical applications, particularly in making parts for vascular treatments.

Already we are seeing a number of companies making medical and medical-related products where RP is a significant part of the manufacturing process. Examples include orthodontic aligners, customized hearing aids, plates for covering cranial holes and small grafts for non-load-bearing bones. The benefit of direct manufacture is the ability to respond quickly, adding value to the overall product design without significant increase in overall cost.

11.5   Tissue Engineering

As can be seen from the examples in this book, there are many people working on the creation of artificial tissue. Skin grafts have been used by surgeons for a number of years to treat burns, for example. Such products do not require advanced manufacturing technology to produce them, relying more on the biochemist than the engineer. Where bony tissue or cartilage is required, the problem becomes much more complex since these would often be required as complex 3D structures and, ideally, should be grown to match the requirements of an individual patient and problem.

Bone, in particular, is a very complex biological system. It takes various forms, with heterogeneous properties that have been adapted through millions of years of evolution to suit very particular requirements. There are currently no advanced manufacturing processes that can match the precision, variation in material properties and all-round complexity of the human body. However, one can see that the use of layer-based manufacturing systems can at least achieve the general requirement of producing a 3D form to be used as a scaffold for growing bony tissue.

Currently, while researchers realize there is a need for heterogeneity in the fabrication methods used, there has been very little success in achieving this. Bones are hollow, with load-bearing directions achieved with the aid of a fibrous structure. Cells must therefore be encouraged to grow in particular regions and in particular directions if they are to become load bearing. One can imagine a time when this may be possible, and indeed some researchers are even thinking beyond this in terms of creating complex cell structures in the form of artificial organs. The key to this is the basic principle of RP, in breaking down a complex 3D problem into a sequence of simpler 2D problems.

11.6 Business

Of course, the bottom line is that all these developments require the input from business and commerce. While medical applications have other driving forces, they are unlikely to be widely used if they are not also commercially viable. Medicine is not just about saving lives but also about improving the quality of life. In many parts of the world, people are willing to pay for such improvements. However, many possible ways to achieve this on a wide scale are infeasible owing to high cost. Advanced manufacturing technology has developed through similar needs to optimize and improve the overall performance of manufacturing to meet the ever increasing demands for more complex products at lower prices. Surely these developments can be put to use in the medical field as well?

Such objectives also incur great risk. It is clear that the technologies mentioned in this book are mostly in very formative stages. The fact that they were developed for manufacturing applications means they are not ideally suited to new application areas like medicine. Lessons have to be learnt, experiments must be performed and careful study is required. Of course this all costs time, effort and money. However, money must be spent, along with time and expertise, if we are to make progress in applying this technology to improving patient care. It must be spent because it is not possible to take risks with human lives. Only where it is known that benefit can be achieved will the technology be applied. Doctors and surgeons must be involved in this process, to ensure that they add their considerable knowledge and expertise to that of the engineer. They must develop the techniques effectively to apply these new technologies to patient care, adding to the considerable human involvement. As well as industrial acceptance, the technologies must undergo the considerable institutional acceptance procedures. Once the medical authorities, food and drug administrations, doctors, surgeons and, of course, patients recognize and approve these technologies for general use, we will be able to see the true benefits of what is described in this book.

Index

3D *see* three-dimensional (3D) reconstruction; three-dimensional (3D) scanning

With kind thanks to Ann Griffiths for creation of this index.